International Encyclopedia of Technical Analysis

Joel G. Siegel, Ph.D., CPA
Jae K. Shim, Ph.D.
Anique Qureshi, Ph.D., CPA, CIA
Jeffrey Brauchler, CPA

Glenlake Publishing Company, Ltd.
Chicago • London • New Delhi

AMACOM
AMERICAN MANAGEMENT ASSOCIATION
New York • Atlanta • Boston • Chicago • Kansas City • San Francisco • Washington, D.C.
Brussels • Mexico City • Tokyo • Toronto

332.632042
I 61

This book is available at a special discount when ordered in bulk quantities.
For information, contact Special Sales Department,
AMACOM, a division of American Management Association, 1601 Broadway,
New York, NY 10019.

This publication is designed to provide accurate and authoritative information in
regard to the subject matter covered. It is sold with the understanding that the
publisher is not engaged in rendering legal, accounting, or other professional service. If legal advice or other expert assistance is required, the services of a com-
petent professional person should be sought.

AMACOM
AMERICAN MANAGEMENT ASSOCIATION
New York • Atlanta • Boston • Chicago • Kansas City • San Francisco • Washington, D.C.
Brussels • Mexico City • Tokyo • Toronto

Dedication

LEOPOLD A. BERNSTEIN
The internationally recognized scholar in investment analysis

ROBERTA M. SIEGEL
Loving wife, colleague, and partner

CHUNG SHIM
Dedicated wife

SHAHEEN QURESHI
Loving and devoted wife

ROBERTA BRAUCHLER
Wonderful wife

Table of Contents

Acknowledgments

We want to express our appreciation to Barbara Evans for the outstanding editorial work she did on this book. We recognize and appreciate her exceptional efforts. Thanks also go to Roberta M. Siegel for her expertise in technical investment analysis.

About The Authors

JOEL G. SIEGEL, Ph.D., CPA, is a consultant in technical investment analysis and professor of accounting and finance at Queens College of the City University of New York. He was previously employed by Coopers and Lybrand and Arthur Andersen. Dr. Siegel has been a consultant in technical analysis to Citicorp, International Telephone & Telegraph, and United Technologies. He is the author of 60 books and about 200 articles on financial topics. His books have been published by The Glenlake Publishing Co., Ltd., International Publishing Corporation, Probus, Richard Irwin, McGraw-Hill, Prentice-Hall, Macmillan, Harper and Row, John Wiley, Barron's, and the American Institute of CPAs. His articles have been published in many financial and investment journals, including the *Financial Analysts Journal, Financial Executive, Journal of Corporate Accounting and Finance, Credit and Financial Management, Long Range Planning*, and *International Journal of Management*. Dr. Siegel has been awarded the Outstanding Educator of America Award. He served as chairperson of the National Oversight Board.

JAE K. SHIM, Ph.D., is an investment consultant specializing in technical analysis models to several companies and professor of accounting and finance at California State University, Long Beach. He received his Ph.D. degree from the University of California at Berkeley. Dr. Shim has 40 books to his credit and has published over 50 articles in financial journals, including *The CPA Journal, Financial Management, International Accountant, Long Range Planning*, and *Decision Sciences*. Dr. Shim was awarded the Credit Research Foundation Award for his article on financial management.

ANIQUE QURESHI, Ph.D., CPA, CIA, is a financial consultant and associate professor of accounting and finance at Queens College. Dr. Qureshi has authored books for Prentice-Hall and has numerous articles appearing in financial journals. He has also made many presentations at conferences.

JEFFREY BRAUCHLER, CPA, a graduate of St. John's University, serves as a financial and investment consultant. He has contributed chapters to several books published by Prentice-Hall. He is the author of numerous articles.

Preface

Technical analysis applies judgment based on experience to help investors predict prices of stocks, bonds, indexes, commodities, futures contracts, and other financial instruments. The focus is on price and volume action. We are concerned as to what has happened and why.

Technical analysis is based on the presumption that history repeats itself. With conditions and factors similar to the past, market participants will basically react in a similar way as previously. The activities of market participants are displayed on charts to identify patterns that repeat over time. Buy and sell signals show up as they have done before. By determining the minimum degree of a trend, one can time just when to buy or sell.

This book covers concepts, terminology, and explanation of technical analysis tools and their practical applications in investment analysis. Technical analysis involves judgment based on experience. The latest developments in technical analysis are included. The book shows the investor what to buy or sell and when. It looks at chart analysis reinforced by technical indicators.

The Commodity Research Bureau (CRB) Futures Price Index measures the trend of commodity prices. The book has a detailed technical analysis of commodities.

The book therefore, as an encyclopedia should, covers the entire spectrum of technical analysis, with enough detail, we hope, to inform the investor new to technical analysis, and refresh the understanding of dedicated chartists.

Introduction

Technical analysis is the study of specific securities and the overall market based on demand/supply relationships. The principles of technical analysis may be applied to any type of security: stock, bond, option, commodity, mutual fund, futures, or indices. A technical indicator may be used to gauge a financial or economic activity.

A technician is a person who uses technical analysis to make investment decisions. The technical analyst bases market forecasting on price movement and other indicators. Technical analysis approaches should be tied to the individual investor's perspective, temperament, personality, and risk profile.

Technicians are of the opinion that a stock will go in the same direction unless there is an interruption from an outside source. They believe the direction and magnitude of the market may be predicted. They attempt to find a consistent price pattern or a relationship between stock price changes and other market information. The relationship is confirmed when two or more technical indicators evidence the same conclusion.

Divergence is the opposite of confirmation. Divergence occurs when technical indicators move in opposite directions, when, for example, one indicator points to increasing stock prices while the other signals declining stock prices. Divergence in an uptrending market implies potential weakness. Divergence in a downtrending market suggests potential strength.

Technical analysis involves many aspects including:

- Charting of price and volume
- Computer software analysis
- Trendline appraisal
- Pattern recognition and analysis
- Study of market breadth

- Mathematical computations, including rates of change, least squares, and moving averages

Several technical indicators may be combined into one model. Each component in the model can be either weighted equally or weighted based on its importance or reliability.

Industries may appear technically strong on their own or relative to the market. Further, companies in an industry can look strong in technical terms relative to the industry and to the overall market.

Many technical market indicators may be used to appraise the overall stock market. The major categories are breadth, monetary, and sentiment. Monetary and interest rate indicators may be useful in predicting future stock prices.

Some indicators are based on subjective assessments. The investor should not rely on only one technical market measure in making an investment decision. Several technical indicators must be used if the decision is to be an informed one.

Technical analysis may be applied to charts showing fundamental data. An example is studying the relationship over time of stock prices and interest rates. A price objective is the technical evaluation of a security's future value. A change in stock price arises from changing investor expectations.

It is important that the investor not buck the underlying trend in the market. An investor must be able to recognize market tops and sell, as well as recognize market bottoms and buy. In other words, the investor must be on guard not to buy a stock at the ending phase of a bull market when a downward turn in stock price is likely or to sell a stock at the last phase of a bear market when an upward trend in stock price is forthcoming.

Technical analysis may improve investment rate of return via better timing of purchases and sales of securities. Timing decisions may also be based on fundamental analysis of the stock.

Most stocks go in the same direction as the overall stock market. If the market is doing well most stocks will also do well, and vice versa. Hence, it is important to assess the condition of the market when formulating an investment strategy.

Probabilities can be used in technical analysis.

Example: The probabilities of stock price at the end of one year follow:

Probability	Expected Stock Price
10%	$50
30%	$45
60%	$48
100%	

The expected stock price equals:

$$10\% (\$50) + 30\% (\$45) + 60\% (\$48) = \$5 + \$13.5 + \$28.8$$
$$= \$47.3$$

In its purest form, technical analysis concentrates on patterns of stock prices and looks at investor moods. Some measures of market moods include trading volume, odd lot transactions, and insider trading.

In looking at movements in price or volume, any time period may be the basis—quarterly, monthly, weekly, daily, and intraday. Though the time horizon may be long-term or short-term, a long-term trend is more revealing for investment purposes.

A primary (major) trend typically lasts more than six months. A minor (near-term) trend typically lasts for two to three weeks and constitutes short-term variability in the intermediate trend. An intermediate-term is typically from two to six months.

A secondary (intermediate) trend, which is a correction in the primary trend, typically lasts for three weeks to three months. Secondary corrections typically retrace one-third to two-thirds of the prior trend.

A correction occurs when security prices go in the opposite direction from a major trend. Many call it a correction when stock prices fall less than 20% but a bear market when they fall 20% or more from the market high.

Stock or commodity prices going up or down will typically decelerate the movement in the direction of the trend and then provide warning signs before changing direction. It is extremely unusual for a trend to reverse instantly.

A one-day reversal is a sudden change in direction of stock price at the end of an intermediate or long-term trend. The change takes place within one trading day.

A change in trend direction may be indicated by the rate of change in price. An important part of technical analysis deals with the study of *momentum* (price velocity). Momentum comes before price. In most cases, where a primary market cycle exists, there is an uptrend in price with greater momentum. However, there is a gradual decline in positive velocity with a decreasing slope of increasing prices.

In almost every case, momentum reaches its peak before the price reaches its top. After that, velocity gradually decreases as price starts to increase a little when a minor and temporary rally occurs. Momentum declines when price rallies occur below prior peaks on minor rally attempts. This is referred to as an exhausted bullish environment.

What happens next is a drastic negative break in momentum as price declines below prior minor lows. This is the declining aspect of the cycle. After an extended period of decline, there is a bottom to price velocity before the price reaches its low. There is less negativity to price velocity on minor drops in price. When there is less negative momentum, a new upward cycle is in the making.

A stock chart shows changes in the volume and price of a specific stock or the overall stock market for a designated period. Charting is more meaningful over a longer period. Technical analysts believe that stock prices move in recurring and identifiable patterns. Irrespective of the periodicity of the data in charts (e.g., quarterly, monthly, weekly, daily, hourly), the principles of technical analysis still apply. The longer the periodicity the easier it is to predict and gain from reviewing stock price changes; the shorter the periodicity, the harder prediction is. To study a chart is to examine the direction of a trend and to assess the likelihood of the trend changing. Chart analysis may be subject to different interpretations depending on the perspective and viewpoint of the chartist. Chart formations reveal useful information to the technical analyst about if and when to buy or sell.

Technical analysts today use computer programs to identify price trends in a market, security, or commodity future. A review of data concerning accumulation and distribution provides useful information about any buildup in buying or selling pressure. Technical analysts (chartists) are not trying to explain why stock prices are increasing or decreasing as they are. Instead, technical analysts are attempting to examine the demand/supply relationship for a stock or bond in order to forecast future prices changes and direction. Technicians want to uncover historical patterns in price and volume as a basis for predicting future prices. As an example, in a *head* and *shoulders* pattern stock price goes up, then down, then up even more than before, and then down again (see Figure 1).

FIGURE 1—ILLUSTRATIVE HEAD AND SHOULDERS PATTERN

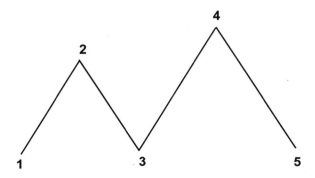

Assuming the historical pattern in Figure 1 continues, Figure 2 shows the anticipated movement in stock price starting at point 5.

FIGURE 2—ILLUSTRATIVE CONTINUATION OF HEAD AND SHOULDERS
                    PATTERN

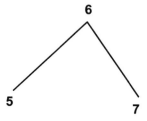

In Figure 2, stock price rebounds from point 5 to point 6, so the stock should be bought at point 5 and sold at point 6 before it decreases to point 7. The stock should then be re-bought at the low price at point 7 because of the expected further rebound in price.

If a company's stock price trend is downward with no indication of a reversal forthcoming, it may be prudent not to buy until there is a technical sign that the declining trend will reverse course. Figure 3 depicts a downward price trend without an indication that the trend is changing.

FIGURE 3—DOWNWARD PRICE TREND OF XYZ COMPANY

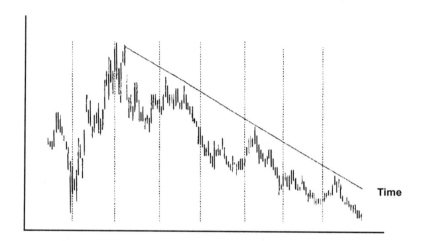

Line studies are technical analysis tools comprising lines drawn on top of a stock's price and/or indicator. They include trendline, support, and resistance.

A determination should be made as to whether prices are *trending* or *trading*. A *trend* refers to market direction. Market moves may be in a series of zigzags. The direction of the peaks and troughs represents the market trend. Lagging indicators should be used during trending markets but leading indicators should be used during trading markets. While it is simple to ascertain if prices are trending or trading, it is very difficult to predict if prices will trend or trade in the future.

Some technical indicators are concerned with confirming a change in trend after it has already taken place; others are concerned with estimating the potential degree or duration of a trend.

By examining area patterns, trends, and trendlines, the technician can identify price direction and any changes thereto. Such an analysis may be useful in ascertaining for short-term trading purposes when to buy or sell. They also provide useful information about when not to sell a long-term position too early.

If the price of a particular stock is "choppy," technical signals about it may not be reliable. However, this volatile stock price situation may offer a greater profit potential, provided a whip saw signal is not evident.

The use and application of moving averages in investing decisions improves the investor's ability to have long-term appreciation in security prices.

The black box approach is a way of doing technical analysis by using mostly mechanical systems in evaluating trends.

Psychological factors have to be considered in determining whether a stock is *over-owned* and subject to a fall in price or *under-owned* and subject to a rise in price. Technical analysts evaluate how people have reacted over time to specified conditions to determine how they will react in the future and the effect on stock price. Such historical patterns may be helpful in identifying a market top or bottom. A premise is that investors will repeat their errors, failing to learn from either history or their own experience. Thus, investors may tend to buy shares when prices are high, and sell shares when prices are low.

In conclusion, technical analysis aids in forecasting stock prices. It studies market action itself: It is a study of prices using primarily chart patterns. The demand/supply relationship and investor expectations are important factors in security prices. The price of a security is its fair value as agreed to by buyers (bulls) and sellers (bears).

Technical analysis improves investment decisions and performance as well as reducing risk. An analysis of historical trends gives clues as to future developments in security prices, including peaks and troughs. A *technical sign* identifies a significant price movement of a security or commodity. If you own a security, would you buy it today? If not, perhaps you should sell it.

A

ACCUMULATION AND DISTRIBUTION

Accumulation is the buying of large amounts of shares in expectation of a significant increase in stock price, often by individuals with deep knowledge of the company who are of the opinion that the stock is undervalued. The accumulation phase is the initial phase of a major upward trend when astute investors will be buying securities. The stock is said to be going from weak hands to strong hands. It may be that bad news has already been discounted by the market. Accumulation takes place when the demand for a security exceeds the supply of it.

An institutional investor may attempt to buy a significant number of shares in a company in a controlled way so as to minimize driving up the stock price. The institutional investor's accumulation plan may take weeks or months to finalize.

The *accumulation area* (see Figure A.1) refers to the price range within which purchasers accumulate shares of a stock. Technicians identify accumulation areas when a security does not drop lower than a specified price. Technical analysts who use the on-balance volume approach to analysis recommend purchasing stocks that have reached their accumulation area, since the stocks can be anticipated to attract more buying interest.

When net volume is constant or rises when prices are dropping, there is accumulation under weakness and a reversal is likely.

Distribution is when a significant amount of shares are sold in expectation of a major decline in stock price. Those doing the selling are sophisticated and knowledgeable about the company. They are of the opinion that the stock is overvalued. Distribution results in lower stock prices on significant volume that is going from strong hands to weak hands.

FIGURE A.1—ACCUMULATION *VERSUS* DISTRIBUTION

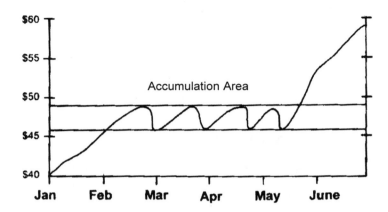

An institutional investor may sell a large block of shares in such a way that the adverse impact on prices is kept to a minimum. Technicians examine a pattern of distribution as an indication that the price of a stock will soon decline.

Constant or declining net volume with a price increase points to distribution under strength and an expected reversal.

Accumulation/distribution is a measure of momentum that compares changes in volume and price. It is assumed that a price move is more significant with increased volume. An increase in accumulation/distribution indicates that the stock is being accumulated (bought), since most of the volume is tied to the increasing movement in price. A decrease in accumulation/distribution means the stock is being distributed (sold), since most of the volume is applicable to the downward movement in prices.

A change in stock price is expected when accumulation or distribution and the stock price diverge. If there is a divergence, prices typically change to be consistent with the accumulation or distribution. Therefore, for instance, if the accumulation/distribution indicator is going down and the stock price is going up, a reversal in price is likely.

Example: ABC Company's stock price diverged when it went to a new high while the accumulation was decreasing. A correction in prices confirmed the indicator (see Figure A.2).

FIGURE A.2—ACCUMULATION VERSUS DISTRIBUTION

The formula for calculating accumulation/distribution is:

Volume x <(close - low) - (high-close)/(high - low)>

A part of the volume for each day increases or decreases the cumulative total. The closer the ending price is to the high for the day, the more is the volume added to the cumulative total; if it close to the low, the more is subtracted. Of course, if the ending price is exactly mid-point between the low and high prices, the cumulative total remains the same.

According to Larry Williams, accumulation (buying) should take place if a stock is going to a new low while the advance-decline ratio fails to make a new low. Distribution (selling) should take place if a stock is going to a new high while the advance-decline ratio does not.

The Williams' accumulation/distribution indicator involves ascertaining the true range low (TRL) and the true range high (TRH).

TRL = Lower of yesterday's close or today's low

TRH = Higher of yesterday's close or today's high

Today's accumulation/distribution is then derived by comparing yesterday's ending price with today's.

1. If today's close is identical to yesterday's close:
   Today's advance-decline (A/D) is 0.

2. If today's close is below yesterday's close:
   Today's A/D = Today's Close - TRH

3. If today's close exceeds yesterday's close:
   Today's A/D = Today's Close - TRL
   Williams' A/D = Today's A/D + Yesterday's Williams' A/D

The *Accumulation Swing Index* developed by Welles Wilder in *New Concepts in Technical Trading Systems* is a cumulative total of the Swing Index.

It attempts to depict a phantom price line of the "real" market considering open, low, high, and closing prices. The index may be evaluated using support/resistance, with consideration given to divergences, new highs and lows, and breakouts. The index reflects a numerical value quantifying price swings, shows near-term swing points, and reveals market direction and strength.

ADVANCE/DECLINE LINE

See Breadth Analysis.

ALTERATION PRINCIPLE

Alteration principle is the theory that the stock market typically does not behave exactly the same twice in a row. For example, if a certain type of top in the market took place last time, it is not likely to do so again.

APEX

Apex is where two lines intersect in a triangle. In the case of a wedge, the apex is characterized by two converging trendlines.

ARBITRAGE

The simultaneous buying and selling of the same or complementary securities, commodities, or currencies in different markets. The arbitrageur buys the security on the exchange with the lower price and simultaneously sells it on the exchange with the higher price. Arbitrage takes advantage of market inefficiencies, while eliminating them.

Example 1: Stock XYZ is trading on the New York Stock Exchange for $5 per share and trading at the same time on the London Exchange for $5.50 per share. A member broker buys 5,000 shares of the stock on the New York Stock Exchange and simultaneously sells 5,000 shares on the London Exchange. The profit is:

($5.50 - $5.00) x 5,000 = $2,500

Example 2: An arbitrageur simultaneously buys one contract of silver in the New York market and sells one contract of silver at a higher price in the Chicago market, making a profit.

Some arbitrageurs buy stock of a firm that may be acquired by another and short the stock of the acquiring entity. If the acquisition takes place, a profit arises; if not, a loss.

ARMS INDEX (TRIN)

The Arms Index (TRIN for Trading Index), developed by Richard W. Arms, Jr., is a short-term trading index that offers the day trader especially a look at how volume—not time—governs stock price changes. It is also commonly referred to by its quote machine symbols, TRIN and MKDS. The Arms Index is designed to measure the relative strength of the volume associated with advancing stocks versus that for declining stocks. If more volume goes into advancing than declining stocks, the Arms Index will fall to a level under 1.00. If more volume flows into declining stocks than advancing stocks, the Arms Index will rise to a level over 1.00.

The index helps signal price changes in market indexes as well as individual issues. You will find Arms indices for the NYSE, the OTC market, the AMEX, and Giant Arms (a combined index for OTC and AMEX). The Bond Arms Index helps forecast interest rates.

The Arms Index is calculated by dividing the ratio of the number of advancing issues to the number of declining issues by the ratio of the volume of advancing issues to the volume of declining issues. It is computed separately for the NYSE, the American Stock Exchange, and NAS-DAQ.

Example: Using the data given in Figure A.3 from "Diaries" in the *Wall Street Journal,*

$$\frac{\dfrac{\text{Advances}}{\text{Declines}}}{\dfrac{\text{Advance Volume}}{\text{Decline Volume}}} = \frac{\dfrac{1{,}130}{775}}{\dfrac{166{,}587}{78{,}016}} = 0.70$$

Figure A.3— DIARIES FOR ARMS INDEX (TRIN)

NYSE	THUR	WED	WK. AGO
Issues traded	2,531	2,547	2,532
Advances	1,130	1,196	597
Declines	775	761	1,412
Unchanged	626	590	523
New highs	85	93	73
New lows	15	24	16
zAdv vol (000)	166,587	238,087	105,901
zDecl vol (000)	78,016	70,598	156,632
zTotal vol (000)	279,550	340,157	289,856
Closing tick	+358	+478	-329
Closing Arms (trin)	.68	.47	.63
ZBlock trades	6,278	7.705	6,490

Source: *Wall Street Journal*

The index is published in *Barron's* and the *Wall Street Journal* and reported daily on TV on *Nightly Business Report, CNNFN,* and *CNBC.* A sample report is presented in Figure A.4.

Figure A.4—ARMS INDEX

ARMS INDEX

The Arms index, also known as the short term trading index, is the average volume of declining issues divided by the average volume of advancing issues. It is computed separately for the NYSE, the American Stock Exchange and Nasdaq. A figure of less than 1.0 indicates more action in rising stocks.

Daily	Feb 8	9	10	11	12
NYSE	.91	1.91	.58	.68	1.16
AMEX	.64	2.96	.43	.19	2.72
NASDAQ	.55	2.17	.97	.23	1.68

Source: *Barron's,* February 15, 1999

A figure of less than 1.0 indicates money flowing into stocks (bullish sign); a ratio higher than 1.0 shows money flowing out of stocks (bearish sign).

One variation of the Arms Index that many technicians monitor is the Open 10 TRIN (also known as the Open 10 Trading Index). It is calculated by dividing the ratio of a 10-day total of advancing issues to a 10-day total of declining issues by a ratio of 10-day total volume of advancing issues to a 10-day total volume of declining issues. A 30-day version of the Open 10 TRIN is also common.

High readings reflect an oversold condition and are generally considered bullish. Low readings reflect an overbought condition and are generally deemed bearish.

Caution: The many studies performed on the Arms Index have come to a variety of conclusions. Many suggest that the Arms Index has relatively limited forecasting value for stock prices.

ASCENDING AND DESCENDING TOPS

Ascending tops is a chart pattern of a stock price over a stated period showing each peak in a stock's price as higher than the previous peak. The upward trend is bullish. (See Figure A.5). Descending tops (Figure A.6) shows each new high price for a security lower than the previous high. The downward trend is bearish.

FIGURE A.5—ASCENDING TOPS

Price

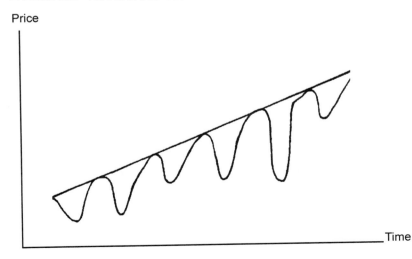

Time

FIGURE A.6—DESCENDING TOPS

Price

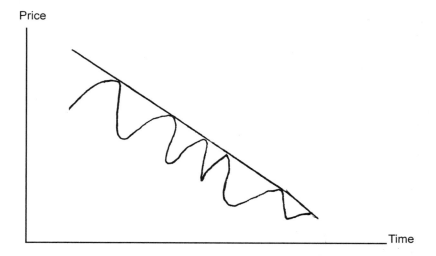

Time

ASK

See Bid and Ask.

ASSUMPTIONS

There are several assumptions associated with technical analysis:

Historical patterns will continue. There is a presumption of repetition in that market participants will react to the market environment the same way as before. Therefore, buying and selling patterns are predictable. The chartist can follow the past to look at the future.

Price trends can be identified on charts. It is presumed that prices will move in a certain direction unless some event prompts a directional change. A price trend will not reverse without an external influence. For example, if stock price is decreasing, it will continue to do so until there is a reversal.

The market reflects demand/supply relationships. Market factors (e.g., interest rate changes, economic conditions, political factors, announced pending acquisitions) are already incorporated into current market price per share. If demand exceeds supply, the stock price will increase, and vice versa. Supply and demand are major factors in determining the real value of a share of stock. It is the market appraisal of prices that governs. It is assumed that there is no value in analyzing financial statements, dividends, profit reports, and similar information in deriving the intrinsic value of a security.

Volume mirrors changes in price.

Figure A.7 shows assumptions applied to an uptrend in overall stock prices.

FIGURE A.7—ASSUMPTIONS

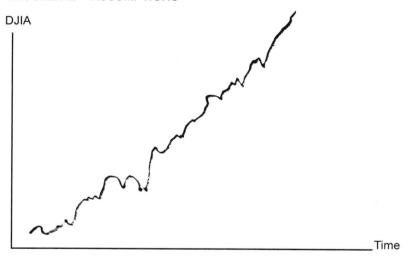

DJIA

Time

AVERAGE TRUE RANGE

See Wilder's Average True Range.

AVERAGE UP AND AVERAGE DOWN

An investor may average up the cost by buying a security or commodity at progressively higher prices. Conversely, the investor may purchase a security or commodity at progressively lower prices to average down the prices paid.

B

BACK UP

To back up is to reverse (turn around) a stock market trend (see Figure B.1).

FIGURE B.1—BACK UP

Price

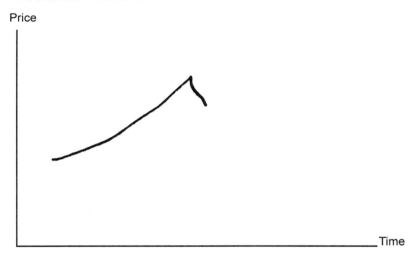

Time

BARRON'S CONFIDENCE INDEX

This index reflects the trading pattern of bond investors to determine timing for buying or selling stock. It is generally presumed that bond traders are more knowledgeable than stock traders, and thus can identify stock market trends earlier. The index is published weekly in *Barron's*:

Yield on *Barron's* 10 top-grade corporate bonds

Yield on Dow Jones 40 bond average = *Barron's* Confidence Index.

Example: The Dow Jones yield is 8% and the Barron's yield is 7%. The Confidence Index is: 7% ÷ 8% = 87.5%.

The numerator will have a lower yield than the denominator because it consists of higher-quality bonds. Because top-quality bonds have lower yields than lower-grade bonds, the index will be below 100%—typically between 80% and 95%. When bond investors are bullish, yield differences between the high-grade bonds and low-grade bonds will be small and the index may be close to 95%.

If confidence is high, investors are likely to buy lower-grade bonds. In consequence, the yield on high-grade bonds will decrease while the yield on low-grade bonds will increase.

An investor who knows what bond traders are doing now may predict what stock traders will do in the future. The lead-time between the Confidence Index and economic conditions and stock market performance is several months.

When bond traders are bullish, an investor may invest in stocks before prices rise. On the other hand, if bond traders are bearish, an investor would not buy stocks or would consider selling current holdings on the expectation that stock prices will fall.

Note: A basis point is a measure of bond yield. One basis point translates into 0.01 percent of yield.

Caution: Bond traders may be making the wrong investment decision, which could result in misleading implications for stock prices. The Confidence Index has a mixed track record in predicting the future. It is deficient in that it considers investors' attitudes only on yields (demand function), ignoring the supply of new bond issues (supply pattern) as it affects yields. A large bond issue by a major corporation, for example, may increase high-grade bond yields, even if prevailing investor attitudes were that yields should be dropping.

BASE

(1) The representative or starting year used as a basis for comparison when doing technical analysis; (2) The vertical line of a triangle formation measuring the height of a pattern.

BASIS POINT

A price level in a chart to aid in formulating the point at which a stop loss should be placed. When technical indicators change, the basing point is adjusted. For example, in a declining market the basis point is adjusted downward.

BASKET TRADE

A significant transaction consisting of numerous shares in different companies.

BEAR MARKET

See Bull and Bear Markets.

BID AND ASK

Bid is the price a market maker is willing to pay for a stock. It is also the price a seller will receive. *Ask* is the price at which the market maker is willing to sell the security. It is also the price the buyer will pay.

BIG MAC INDEX

McDonald's Corporation's Big Mac sandwich is the basis for the Big Mac Index that helps explain global foreign exchange rates. Comparing the cost of a sandwich at different prices worldwide at current exchange rates ignores the impact of foreign central bank intervention and trade deficits. A comparison of the cost of a Big Mac in New York with overseas prices indicates the strength or weakness of foreign currencies against the dollar.

BLOWOFF

An occurrence at a major market top in which prices and trading abruptly rally significantly after a long advance. In futures trading, it is associated with a significant decline in the open interest.

BOLLINGER BANDS

Named after John Bollinger, Bollinger Bands are similar to trading bands (also called moving average envelopes). However, while trading bands are plotted at a fixed percentage above and below a moving average, Bollinger Bands are plotted at standard deviation levels above and below a moving average.

There are problems with trading bands: (1) different trading situations require different widths, and (2) different moving average lengths produce different trading bands. Bollinger solved these problems by placing the bands two standard deviations on either side of the moving average. Bollinger Bands will therefore vary in distance from the average as a function of the stock's volatility. Since standard deviation is a measure of volatility, Bollinger Bands are self-adjusting, widening during volatile markets and contracting during calmer periods.

Bollinger Bands displayed on prices are usually displayed on top of the prices, but they can be displayed on an indicator. As with moving average envelopes, prices tend to stay within the upper and lower Bollinger Bands. The distinctive characteristic of Bollinger Bands is that the spacing between the bands varies based on the volatility of the prices. During periods of extreme price changes (i.e., high volatility), the bands widen to become more forgiving. During periods of stagnant pricing (i.e., low volatility), the bands narrow to contain prices.

According to John Bollinger:

- Sharp moves tend to occur after the bands tighten to the average (i.e., volatility lessens).

- A move outside the bands calls for a continuation of the trend.

- Tops and bottoms made outside the bands that are followed by tops and bottoms inside the bands indicate a trend reversal.

- A move originating at one band tends to go to the other band.

Example: Figure B.2 shows Bollinger Bands on Microsoft prices that were calculated using a 20-day exponential moving average; they are spaced two deviations apart. The bands were at their widest when prices were volatile during July. They narrowed when prices entered a consolidation period later in the year. Narrowing of the bands increases the probability of a sharp breakout in prices. The longer prices remain within the narrow bands the more likely is a price breakout.

B

FIGURE B.2—BOLLINGER BANDS AND MICROSOFT

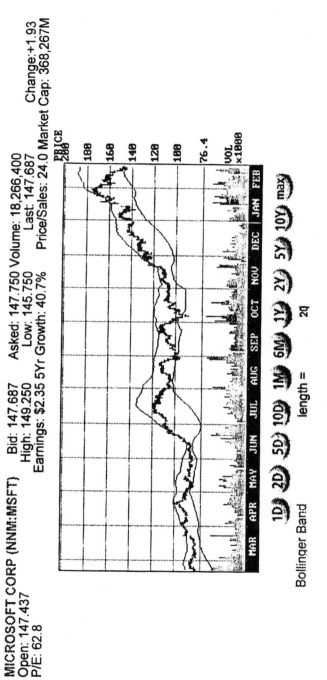

BOTTOM

This is the support level for a stock or market price. If a stock price goes below the support level, it is said that the "bottom dropped out." If prices trend upward again, it is said that the stock has "bottomed out." A *bottom fisher* is an investor who buys a stock that has fallen steeply below the support level. Identifying high quality companies before taking into account the overall economic picture and trend is known as the bottom-up approach to investing.

BREADTH ANALYSIS

Breadth is the extent of investor participation in stocks, bonds, and commodities. It applies to the dispersion of a general price increase or decrease. Breadth indicators track stock, bond, and commodity market trends in order to assess market strength and weakness. They show how many issues are participating in a market move.

The trend in market participation is important. In general, breadth is good if two-thirds of the stocks on an exchange are increasing. When breadth is good, the market is likely to last longer, since more investors are participating.

With regard to stock, breadth is the net difference between advancing issues and declining ones. The number of advancing stocks are those traded on an exchange (e.g., New York Stock Exchange) that went up in price on that trading day; declining stocks are those that went down. The market is bullish when advancing issues are significantly more than declining issues, bearish if declines outnumber advances.

The fewer issues going in the same direction as the market averages, the more apt there is to be a trend reversal. Further, the longer a price trend is maintained without a follow-up in the overall market, the more vulnerable is the advance.

Breadth relates to the number of sectors (e.g., industry groups such as airline, oil, autos, or technology) with market participation. If many sectors are enjoying an advance, this is a bullish indication. However, if only one or a few market sectors are involved in the advance, this is a bearish sign. In other words, breadth in the market reveals the extent to which a market index is supported by its components.

Market breadth shows whether the environment for stocks is good or bad. Breadth indicators such as positive and negative divergences signal major turning points in the market. Breadth divergences should be confirmed by a trend reversal in the market indices. Market breadth is an advance indicator of major stock price advances or declines.

When the difference between the number of advancing and declining issues is low, there is a market standoff that probably will lead to fairly stable prices. An extremely high number of advancing issues, particularly based on a 10-day moving average, are very bullish. An extremely high number of declining issues are very bearish.

The broad market usually leads the market averages at market tops, for example, when the broad list of stocks peaks out ahead of a market average like the Standard & Poor's 500. Breadth analysis is not useful to ascertain major reversals at market bottoms because most stocks coincide with or lag behind the major market averages.

In a thin (narrow, inactive) market where trading for securities or commodities is light, resulting in wide variability in prices compared to volume, there are few bids to buy and few offers to sell. Liquidity is a problem. Institutional investors usually avoid narrow markets because it is difficult to get in or out of a position without significantly affecting the price of the security.

On the other hand, in an active (tight) market a stock may experience active trading and bid-offer price spreads narrow. The most active issues are those that have the largest share volume; these are very marketable. Such stocks account for about 20% of total NYSE volume. Summaries of the most active stocks are prepared by technical services such as *Indicator Digest*. The investor must consider the degree of price changes as well as heavy trading activity.

Breadth indicates market momentum, an increasing pattern of trading over time. In other words, volume starts an increasing trend. Momentum relates to the degree of change in the volume or price of a security or securities in the overall market over a specified period.

Breadth thus applies to how many issues is part of a move upward or downward. When the number of advancing stocks decreases as the rally unfolds, the rally is questionable. Similarly, the indicator is bullish when a declining market involves fewer stocks in price over time.

ABSOLUTE BREADTH INDEX

The absolute breadth index (ABI) formulated by Norman Fosback equals advancing issues less declining issues. Absolute value is irrespective of sign; therefore, an absolute value of +60 or -60 is expressed as just 60.

Example: On a given trading day advancing issues on the New York Stock Exchange were 1,237 and declining ones were 816. The net difference (the ABI) is 421.

By plotting advancing and declining issues, one can determine the strength or weakness in the market. If advancing issues substantially exceed declining issues, strength exists in the market; weakness is indicated in the converse situation. Although daily values themselves may be plotted, a 10 to 60-day exponentially moving average would be more insightful in evaluating an overbought or oversold situation.

The ABI reflects the degree of activity, fluctuation, and change on a stock exchange without considering price direction. For example, a high reading represents an active and changing market. A low reading represents a lack of change.

If the absolute difference between advancing and declining issues is high, a market trough is more probable than a market peak. In other words, a market bottom may be close at hand when the index is high because a buying spree typically takes place near a market bottom. A higher index usually signals higher prices within three to 12 months. A low ABI more likely means a slow topping situation usually indicating a market top, where stock prices will most likely decline.

In determining ABI we usually use weekly or daily NYSE information. A variation of ABI is: Modified ABI = Weekly ABI divided by total issues traded. The modified ABI should be subject to a 10-week moving average. A reading above 40% is quite bullish, a reading less than 15% quite bearish.

ADVANCE/DECLINE DIVERGENCE OSCILLATOR

The advance/decline divergence oscillator was developed by Arthur Merrill. If the advance/decline line is decreasing but the market index is increasing, the divergence creates the premise that the market index will top and reverse its trend. The advance/decline divergence oscillator is computed in the following steps:

1. Total the daily number of issues advancing, declining, and unchanged on the NYSE for each week of the previous year.

2. Calculate Edmund Tabell's weekly ratio:

$$\frac{\text{Number of advancing issues - number of declining issues}}{\text{Number of unchanged issues}}$$

This ratio reveals market conviction. A high ratio indicates a low number of unchanged stocks, indicating a high market conviction; the reverse signals low market conviction.

3. Add the values of the ratio for the previous 52 weeks.

4. Using regression analysis, plot a regression line based on computed data and weekly closing price of the DJIA for the previous 52 weeks. The X-axis will show the cumulative ratio while the Y axis presents the DJIA. In ascertaining the DJIA's expected value for a particular value of the cumulative ratio, identify the point on the regression line directly above the cumulative ratio amount (see Figure B.3).

FIGURE B.3—ADVANCE/DECLINE DIVERGENCE OSCILLATOR (ADDO)

DJIA

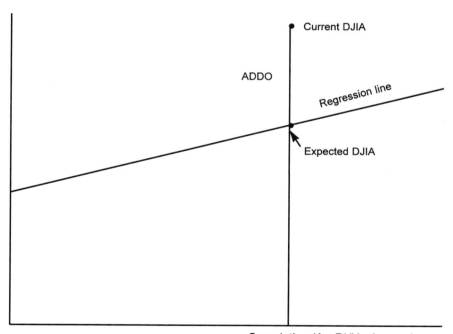

Cumulative (A - D)/Unchanged

5. Compute the advance/decline divergence oscillator as the percentage deviation of the DJIA's current closing price from its expected value. A positive oscillator reading implies that the price of stocks is increasing faster than anticipated by the advance/decline line and a downturn is probable; with a negative oscillator, price is increasing more slowly than anticipated, so an upturn is likely.

ADVANCE/DECLINE LINE

An important measure of market breadth is the advance/decline line measuring the breadth of an advance or decline in a particular stock. It represents a cumulative total of the advancing/declining issues. The advance/decline line is a good measure of strength or weakness in the market when it is compared to the change in a stock market index (e.g., Standard and Poor's 100).

The advance/decline line trends upward when more stocks are advancing, downward when they are declining. It indicates whether the stock market is in an upward or downward trend, if the market is solid, and the length of time of the present trend. The trend of the advance/decline line is more important than its value or level.

The advance/decline line equals:

Cumulative total + (Number of stocks increasing in price for the day—Number of stocks decreasing in price for the day)

The calculation is usually based on weekly or daily NYSE data but may be used to evaluate issues on the American Stock Exchange and over-the-counter issues.

Example: At the start of the trading day, the cumulative total of the advance/decline line was 210. During the day, advancing issues were 1,330 and declining issues 1,105. The cumulative total at the end of the trading day:

$$210 + (1,330 - 1,105) = 435$$

A divergence between the advance-decline line and a stock market index (Standard and Poor's 500) is very revealing. For example, if the index is attempting a new high while the advance/decline line is starting to move down, the bull market may be at its end with a correction in stock prices imminent.

The longer and greater the divergence between breadth and prices, the more significant a decline is likely to be. As a result, divergence between the advance/decline line and a market index at primary peaks is more important than at intermediate tops. However, absence of a divergence does not necessarily suggest that a significant bear market may not occur.

A positive divergence occurs at market bottoms when the advance/decline line does not confirm a new low in the Dow Jones Industrial Average. In most cases at market bottoms the advance/decline line either coincides with or lags behind the DJIA low. There is no predictive import until a reversal in downtrend is indicated by a breakout from a trend line, price pattern, or moving average crossover.

Since daily advance/decline lines tend downward, care should be taken in relating recent highs with those accomplished two to three years previous. Daily advance/decline lines come into their own if they do not confirm new highs in the market average within the last 18 months.

Some bottoms in the daily line typically coincide with or lag after market average bottoms; they are not particularly useful to spot a trend reversal.

The advance-decline line may be analyzed at any time. It is an accumulation measure of stock market breadth that starts with a base value (e.g., 10,000, 50,000). Advances are then added and declines are subtracted.

Example:

Day	Advancing Issues	Declining Issues	Change	Value of Advance/Decline Line
0				50,000
Mon.	960	800	+160	50,160
Tues.	700	1,000	-300	49,860
Wed.	1,210	600	+610	50,470

A drawback in advance/decline line analysis is the objectivity required to evaluate it. Comparative relationships may be misinterpreted. Table B.1 shows the advance/decline line can be analyzed relative to an overall market index such as the Standard & Poor's 100.

Because the DJIA and other market indexes consist of larger, financially stronger, companies, the indexes tend to continue to advance well after the overall market has peaked.

TABLE B.1—BREADTH ANALYSIS: COMPARATIVE ANALYSIS OF THE ADVANCE/DECLINE LINE

Advance/Decline Line	Stock Market Index	Conclusion
Increasing	Decreasing	Bullish
Decreasing	Increasing	Bearish
Materially below corresponding top	Close to or at prior top	Bearish
Materially above corresponding top	Close to or at prior top	Bullish
Materially above previous bottom	Close to or at prior bottom	Bullish
Materially below previous bottom	Close to or at prior bottom	Bearish

The advance/decline line usually increases and decreases consistently with the major market indexes but typically peaks well before the top of a market average, for the following reasons:

1. Lower-quality stocks provide the possibility for the greatest return but also have the highest risk and the greatest losses. Since blue chip stocks are of the highest quality, they are the last stocks to be sold in a bull market.

2. The market generally discounts the business cycle and usually has a bullish market peak six to nine months before the economy tops. Because a peak in business activity comes after a worsening in some leading indicators, stocks representing certain sectors (e.g., financial, construction, consumer spending) will peak before the overall market.

3. Most stocks on the NYSE are sensitive to interest rate changes. Because interest rates increase before market peaks, interest-sensitive stocks will decline.

When analyzing advance/decline information, consider the following:

1. Breadth data may diverge negatively from market indexes. However, a key rally is often indicated when there is a down trend-line violation with a breakout in the market average.

2. In most situations weekly advance/decline lines have less downward bias than lines constructed with daily data.

3. It is expected that the advance/decline line will coincide or lag at market bottoms. If the advance-decline line does not confirm a new low in the index, it is unusual and favorable, but only if confirmed by a reversal in the average.

4. The relationship between the advance/decline line and a market index for an extended time should be examined to ascertain any bias.

5. A discrepancy between the advance/decline line and a market index at a market top is typically cleared up by a decline in the average. However, there must be a trend reversal sign before it can be concluded that the average will also decline.

The steepness of the advance/decline line (Figure B.4) graphically represents whether there is a strong bull or bear market.

FIGURE B.4—BREADTH ANALYSIS: ADVANCE/DECLINE LINE

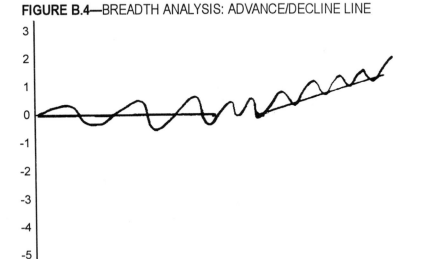

ADVANCE/DECLINE RATIO

The advance/decline ratio is the result of dividing the number of advancing issues by the number of declining ones, based on weekly or daily NYSE figures. It indicates market strength or weakness, and whether the trend will continue.

Example: If stocks advancing in price for the day numbered 1,215 and those declining numbered 860, the advance/decline ratio would be 1.4.

The ratio is both a momentum indicator and a reflection of whether the market is over or undervalued. The investor may compute a moving average (for smoothing purposes) of the advance/decline ratio to indicate whether the market is overbought or oversold. A high advance/decline ratio (over 1.25) may imply an overbought scenario and a possible correction. A low ratio (below .75) may imply an oversold market and a forthcoming rally.

Figure B.5 shows overbought/oversold situations based on the advance/decline ratio analysis.

FIGURE B.5—ADVANCE/DECLINE RATIO ANALYSIS

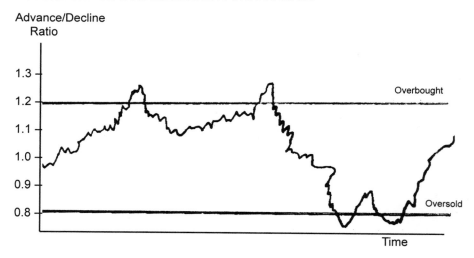

BOLTON-TREMLAY INDICATOR

The Bolton-Tremlay Indicator represents a cumulative advance/decline measure based on the number of unchanged issues. It is computed in five steps:

1. Find the ratio of the number of advancing issues divided by the number of unchanged issues.
2. Find the ratio of the number of declining issues divided by the number of unchanged issues.
3. Subtract the declining ratio (step 2) from the advancing ratio (step 1).
4. Take the square root of the difference.
5. Add the square root (step 4) to the square root of the previous day's Bolton-Tremblay Indicator. Retain the sign, a plus for net advances or a minus for net declines.

Example: Advancing issues are 1,300, declining ones are 400, and unchanged issues are 100. Today's Bolton Tremblay's Indicator equals:

Yesterday's Bolton-Tremblay Indicator + Square Root of (1,300/100 - 400/100)
Yesterday's Bolton-Tremblay Indicator + Square Root of 9
Yesterday's Bolton-Tremblay Indicator + 3

The index may be computed at any time, but a positive beginning value such as +500 should be used to avoid plotting negative numbers on the chart.

The Bolton-Tremblay Indicator is plotted and interpreted like the advance/decline line. The investor should concentrate on divergences between the indicator and an overall market index such as the Standard & Poor's 500 so as to assess market strength or weakness and changes in trend direction. Trend is more important than actual value in this measure.

BREADTH THRUST

A breadth thrust indicator developed by Martin Zweig may be calculated as a measure of momentum in the stock market. It is a 10-day simple moving average of the following ratio:

Advancing issues divided by (advancing issues + declining issues)

Neutral issues are ignored in this calculation.

NYSE information is used in the computation.

A high reading (.55 or greater) is a bullish indicator while a low reading (.45 or lower) is bearish.

There is a breadth thrust when the ratio goes from less than 40% to more than 61.5% over a 10-day period. Many bull markets start with a thrust. A thrust implies that the securities market is moving very quickly from being undervalued to one of strength without yet being overvalued (see Figure B.6).

FIGURE B.6—BREATH THRUST ANALYSIS

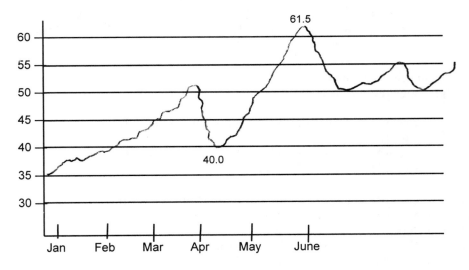

HAURLAN INDEX

The Haurlan Index was introduced by P.N. Haurlan. It has the following three ingredients:

1. A long-term component consisting of a 200-day exponential moving average of the difference between advancing and declining issues on the NYSE.

2. An intermediate-term component consisting of a 20-day exponential moving average of the difference between advancing and declining issues on the NYSE.

3. A short-term component consisting of a 3-day exponential moving average of the difference between advancing and declining issues on the NYSE.

The long-term element is used to identify a major trend in stock prices. The intermediate-term component provides buy and sell signals when trend lines or support and resistance levels are crossed. If the short-term element goes higher than +100, a short-term buy is warranted. Stocks bought should be sold when the index drops to -150.

HUGHES BREADTH INDEX

The Hughes Breadth Index is an advance/decline non-cumulative indicator. It equals (number of advancing issues - number of declining issues) divided by the total number of issues traded.

The indicator is based on weekly or daily NYSE information. The premise underlying the Hughes Breadth Index is that an improvement in market aspects results in rising stock prices, and a deterioration results in declining prices. This index is designed to appraise market strength or weakness. Breadth analysis emphasizes change instead of level.

Example: On a particular day, net declining issues are 58. Securities traded are 1,475. The breadth index equals 3.9.

It is best to use a 10-day simple moving average. A reading above +.08 is bullish and a reading below -.2 is bearish. The higher the plus percentage, the more positive the market, because more stocks are increasing in price relative to those decreasing in price.

Breadth indexes for a number of years may be compared. The breadth index may also be compared to a base year or included in a 150-day moving average.

McCLELLAN OSCILLATOR

The McClellan Oscillator is a near-term to intermediate-term market breadth indicator using NYSE information. The McClellan Oscillator equals (the 19-day exponential moving average of the difference between the number of advancing issues and the number of declining issues) - (the

39-day exponential moving average of the difference between the number of advancing issues and the number of declining issues)

The McClellan Oscillator is used to determine if the market is overvalued or undervalued. The oscillator will usually have an extreme reading before a change in the trend in stock price. A reading for excessive buying, indicating an overvalued market, is one over 100. A reading of about -150 reflects a bear market selling climax.

The McClellan Oscillator typically goes through zero near market tops and bottoms. If the oscillator goes from above to below zero, it is a bearish sign; movement upward is bullish.

The McClellan Summation Index, a cumulative total of the McClellan Oscillator, helps to identify intermediate to long-term stock market moves. It is similar to the McClellan Oscillator in that it provides buy and sell signals when crossing zero.

OPEN TRADING INDEX

A variation of the Arms Index is the Open Trading Index (Open - 10 TRIM), which measures market breadth (strength) by looking at both advancing/declining volume and advancing/declining issues. An Open-10 TRIM indicator of less than 0.90 is a bullish sign while one of more than 0.90 is bearish. The index is computed by tracking a 10-day total of each TRIM component before making the TRIM:

Open Trading Index = <10-period total of advancing issues divided by 10-day period total of declining issues> divided by <10-period total of advancing volume divided by 10-period total of declining volume>

When the stock market is significantly increasing, the Arms Index decreases to low levels (0.5 or less) because the up/down volume ratio is typically much larger than the advance/decline ratio. When the stock market is down, the Arms Index is typically more than 1.0 due to heavy down volume.

OVERBOUGHT/OVERSOLD INDICATOR

Another market breadth measure is the overbought/oversold indicator, which equals the 10-period moving average of advancing issues less declining issues. It indicates when stocks are overvalued with a correction likely and when they are undervalued with a rally likely. A reading above +200 is bearish while a reading below -200 is bullish. If the overbought/oversold indicator goes below +200, it is time to sell. When the indicator goes above -200, it is time to buy.

SCHULTZ'S RATIO

Schultz's advances/total issues traded measure equals the number of advancing issues divided by the total number of issued traded.

Schultz's ratio is typically based on daily information from the NYSE. A simple moving average (e.g., 15-days) may be used based on daily figures so as to smooth out fluctuating movements. A very high (low) Schultz's ratio infers a slight bullish (bearish) sign. The measure, however, should not be used for market timing decisions.

ARMS INDEX

The Arms Index (see above) can also be used in breadth analysis. As a general rule, when the Arms Index is less (more) than 1.0 it is bullish (bearish). The Index is also helpful in identifying an overbought/oversold market. If the Index points to an overbought market, sell; to an oversold market, buy (see Figure B.7).

FIGURE B.7—ARMS INDEX ANALYSIS

Arms Index

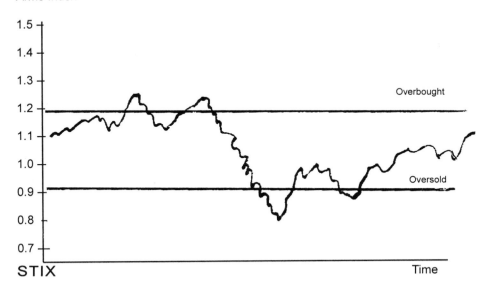

STIX

STIX represents an exponentially smoothed near-term measure of market breadth; it is computed daily in stages based on advance/decline information on the NYSE. The steps are:

1. Advance to total stocks changing in price ratio (ATSC) equals the number of stocks advancing divided by the number of stocks advancing plus the number of stocks declining.
2. Adjusted ATSC = ATSC x 0.09
3. Adjusted ATSC + Previous day's STIX value x 0.91
   The smoothing is for a 21-day simple moving average.
   A high STIX reading is bullish, a low reading bearish.

25-DAY PLURALITY INDEX

The 25-day Plurality Index is computed by *subtracting* the difference between the number of advancing and declining issues *and dividing by the last* 25 trading days. If the index goes above 12,000, it is bullish; if it goes below 6,000, it is bearish. NYSE data is used.

UNCHANGED ISSUES INDEX

The unchanged issues index equals the number of unchanged issues divided by the total number of issues traded. The computation is based on daily or weekly NYSE data. It is presumed that stock prices form tops over extended time periods but quickly bottom out. Hence, the ratio should be high before a market top and low before a market bottom.

UPSIDE-DOWNSIDE INDICATOR

We can similarly calculate the upside-downside indicator by subtracting the daily volume of declining stocks from the daily volume of advancing stocks.

This indicator reflects the new flow of volume into or out of the market. For example, a +10 reading shows that up volume exceeds down volume by 10 million shares. A -10 reading reveals that down volume exceeds up volume by 10 million shares.

We can use this indicator to compare volume over a number of days. A typical reading is about +-50. An extremely active day might be +-300 million shares.

Martin Zweig's Breadth Advance/Decline Indicator, published in the *Zweig Forecast* and reported daily on CNBC, is used for stocks traded on the New York Stock Exchange. It is similar to the customary advance/decline ratio except it uses a 10-day moving average of the number of advancing issues to declining issues.

UPSIDE/DOWNSIDE RATIO

The upside/downside ratio tracks how up (advancing) volume compares to down (declining) volume on a particular stock exchange. The ratio equals daily volume of advancing stocks divided by daily volume of declining stocks.

This ratio reveals buying and selling pressure. A high reading signals buying pressure, which is bullish, and a low reading signifies selling pressure, which is bearish.

An upside/downside ratio that exceeds 1.0, means that more volume applies to stocks going up in price than going down. A very high ratio (e.g., 8 to 10) is encouraging, particularly if it occurs more than once in a short period of time. According to Martin Zweig, a ratio exceeding 9 is very bullish.

BREADTH (ADVANCE-DECLINE) INDEX

The Breadth (Advance-Decline) Index computes for each trading day the net advances or declines in stocks on the NYSE. A strong market exists when there are net advances; a weak market exists when there are net declines. Of course, the magnitude of strength depends on how much advancing issues outnumber declining ones. On the other hand, if declining stocks outnumber advancing stocks by 3 to 1, market weakness is significant. The index is computed:

$$\text{Breadth Index} = \frac{\text{Number of net advances or declines in securities}}{\text{Number of securities traded}}$$

Example: Net advancing issues are 230. Securities traded are 2,145. The Breadth Index equals:

$$\frac{230}{2,145} = +0.107$$

The higher the plus percentage the better, since that means more stocks are increasing in price relative to those decreasing.

The Zweig version of the Breadth Advance/Decline Indicator equals

$$\frac{\text{10-day moving average of advancing issues}}{\text{10-day moving average of declining issues}}$$

The Breadth Index may be computed easily by referring to the financial pages of a newspaper; the market diary section will provide the number of advancing and declining issues, along with the number of issues unchanged, as well as the total number of issues traded, which is the sum of these. Some financial advisory publications calculate the index, relieving the investor from performing the computation. Martin Zweig's version is published in the *Zweig Forecast*. The financial news programs on CNBC report it daily. Figure B.8 shows the totals for market advances and declines as published in *Barron's*. Figure B.9 shows the NYSE Composite Daily Breadth. Figure B.10 shows a bond diary, including advances and declines, from *Barron's*.

FIGURE B.8—MARKET ADVANCE/DECLINE VOLUME

TRADING DIARY
Supplied by "Quotron, QCHA" is the average
percentage movement for all exchange listed
stocks each day on an unweighted basis.
Market Advance/Decline Volumes

Daily	Feb 8	9	10	11	12
NY Up	327,254	141,909	364,009	523,082	157,826
NY Off	339,901	540,839	331,618	239,710	492,494
% (QCHA)	-.05	-.82	-.48	+.53	-.91
Amex Up	18,344	4,646	13,805	27,175	4,767
Amex Off	10,073	25,118	10,164	4,538	21,720
% (QCHA)	+.53	-.87	-.77	+.45	-.26
NASD Up	516,477	143,660	353,983	796,715	200,891
NASD Off	327,873	728,448	521,672	126,670	602,884
% (QCHA)	+.38	-1.44	-.24	+1.47	-.58

Source: *Barron's*, Feb. 15, 1999

FIGURE B.9—NYSE COMPOSITE DAILY BREADTH

Daily	Feb 8	9	10	11	12
Issues Traded	3,581	3,574	3,554	3,572	3,555
Advances	1,403	1,014	1,177	1,787	830
Declines	1,595	2,027	1,855	1,202	2,228
Unchanges	583	533	522	583	497
New Highs	21	16	19	20	14
New Lows	73	81	97	104	161
Blocks	14,007	14,196	14,141	16,633	13,097
Total (000)	857,225	857,653	861,914	949,456	816,099

Source: *Barron's* Feb. 15, 1999

FIGURE B.10—NYSE BOND DIARY

	2/08	2/09	2/10	2/11	2/12
Total	216	223	225	200	227
Advances	85	87	87	60	60
Declines	96	91	101	97	115
Unchanged	35	45	37	43	52
NewHighs	2	2	4	2	1
NewLows	8	6	6	8	5
Salesth$	12,313	16,036	14,338	13,590	12,788

Source: *Barron's* Feb. 15, 1999

Breadth analysis emphasizes change rather than level. The Breadth Index should be compared to popular market averages. Typically, they are consistent; in a bull market, an investor should be on guard against an extended disparity, as for example when the Breadth Index gradually moves downward to new lows while the Standard & Poor's 500 Index reaches new highs.

Breadth Indexes over a five- to ten-year period can be compared. The Breadth Index may be compared to a base year or included in a 150-day moving average.

The investor uses market direction to identify strength or weakness. Advances and declines usually move in the same direction as standard market averages (e.g., the Standard & Poor's 500 Index and the Dow Jones Industrial Average). However, they may go in the opposite direction at a market peak or bottom.

The investor can be confident of market strength when the Breadth Index and a standard market index are increasing, and may buy securities, since a bull market is indicated. Decreasing indexes indicate market weakness, and a good time to sell.

Caution: Historically, stock advances have exceeded declines. However, a sudden reversal can occur, causing a net decline.

BREAKOUT

When the price of a security or commodity moves out of its previous range, we have a breakout. A false breakout may occur in which the breakout quickly reverses itself and the price of the security or commodity then goes in the opposite direction. A premature breakout is a temporary, short-lived breakout with a quick retreat back into the pattern.

However, it will ultimately break out again without retreat and in fact continue its movement.

BULL SPREAD

An option strategy executed with calls or puts. The investor earns a profit as the underlying security increases in value. A vertical spread occurs when there is a simultaneous purchase and sale of options on the same type of security at different strike prices but with the same expiration date. A calendar spread is the simultaneous purchase and sale of options of the same class and the identical price but with different expiration dates. A diagonal spread is a combination of vertical and calendar spreads: the purchase and sale of options of the same class but with different strike prices and different expiration dates.

BULL AND BEAR MARKETS

A bull market is one in which most securities are increasing in price on most days. In a bear market, most stocks are decreasing in price most of the time. Most investors make money in a bull market and lose money in a bear market. There has been a recurrence over time in the regular pattern of major bull and bear markets. In general, about 75% of all stocks move with the general market.

In selecting stocks or commodities, bullish or bearish implications in the market have to be considered. For example, in a bear market, there is greater likelihood that the stock selected will have a downward bias.

In the stock market cycle, investors typically go from pessimism and occasionally panic at market bottoms to over-optimism and occasionally euphoria at market tops. Any reversal in trend should be identified early.

A high ratio of a low-priced index to a high-priced index means more speculation in the market. A high level of speculation often signals the last stage of a bull market. At market bottoms, the low-priced stock index either lags or coincides with the Dow Jones Industrial Average. It is a confirming indicator. At market tops, the low-priced stock index typically leads by as much as 15 months.

A double bottom is a pattern in which stock price has experienced one bottom and then declines again to about the same level, and the stock price increases again. A double top is the converse: the price peaks twice and then falls.

A triple top is a pattern showing that stock price has experienced a double top increasing again to about the same level and then falling again. In a triple bottom, the stock price hits about the same low three times with brief rises between before beginning to increase.

A rounding top is an upside-down saucer-shaped curve occurring when an uptrending stock price gradually alters into a downtrend over a stated time period (e.g., a few months). When a downtrending curve gradually rounds through a bottom and trends up, it is considered a rounding bottom.

A V-shaped bottom is a price pattern in which there is a sharp downtrend that abruptly changes to a sharp uptrend.

SURVEYING INVESTORS

Attempting to determine whether market watchers and players are likely to be buyers (bulls) or sellers (bears) of stock in the near term, various groups conduct polls of various sets of investors, ranging from professional traders to the small stockholder. Typically, they ask one key question: Are you bullish on the stock market? The survey than adds up the percentage of those polled who say they are bullish. A bullish reading can be anything over 50% or, after a prolonged market drop, a turn from increasing bearishness to increasing bullishness.

Investment advisors are surveyed weekly by *Investors Intelligence* of New Rochelle, New York, to determine if they are neutral, bullish, or bearish on the securities market. Investors Intelligence publishes the results as the bull/bear ratio. This market sentiment indicator is used as a contrarian indicator in making investment decisions: If the ratio is 75% or more, investment advisors are extremely bullish, a sign of a possible market top. Conversely, if the bull/bear ratio is 25% or lower (advisers are bearish), this may signal a market bottom. Thus, if investment advisors are very bullish, sell. If they are very bearish, buy.

In a falling market or one near a possible bottom, the slightest turn to bullishness could be a buying signal. At such a low point for stocks, watch for any hint that bearish sentiment is ebbing.

In calculating the bull/bear ratio, we ignore investment advisors who are neutral. The ratio equals:

Bullish advisors
Total bullish and bearish advisors

Another form of the ratio that may be useful to examine is:

Bullish advisors
Bearish advisors

Bullish consensus is based on a weekly survey of newsletter writers carried out by Hadady Publications (Pasadena, California). If the reading is less than 30% bullish, the market is considered undervalued and offers a buying opportunity; if 80% of newsletter writers are optimistic, the market is deemed overvalued and susceptible to a decline in prices.

BULLS

A bull market is one in which for a period of time, typically from one to several years, stock prices move upward. Any downturn in prices is short-lived. A strong bull market can occur only as long as buying pressures continue to be strong.

Investors Intelligence issues each week the bullish sentiment of stock market professionals. It is a contrary indicator: If 55% are bullish the market is overvalued; if 35% are bullish, the market is undervalued.

A bull market is characterized by:

1. Accumulation.
2. Constantly increasing stock prices with graduated volume.
3. Activity by average investors buying a significant volume of shares. The investor who has been on the sidelines may attempt to get in the market at this point.

Caution is dictated when stock prices are at historically high levels because most bull markets in their final phases are more risky.

A *buying climax* refers to a rapid increase in security or commodity prices, setting the stage for a fast fallback. The surge attracts many potential buyers to a security, making them vulnerable because there are few to sell their shares to at the higher prices. This is what prompts the ensuing fall. A buying climax is a dramatic run-up coupled with increased trading volume.

A *bull trap* occurs when prices break out to the upside but then turn down again, causing losses. In a bull trap there is a false indicator when a declining trend in a market index or stock has reversed.

BEARS

A bear market is one experiencing a decline of 20% or more after reaching a peak. A bear market typically lasts six months to several years, during which stock prices keep declining. Any upward movement in price is brief.

There are usually three phases in a bear market:

1. *Distribution.* This in essence starts with the last phase of a bull market, when farsighted investors consider the market to be fundamentally overvalued and accelerate their selling of shares. While there is active trading volume, it diminishes on rallies. The investing public, while still active, is becoming more cautious.
2. *Panic.* Buyers thin out and sellers are become very scared. There is significant downward pressure in stock prices with a huge jump in volume.
3. *Washout,* involving the sale of positions by investors who have held out up to this point.

In a bear market a bottom, the lowest point in a sequence of prices trending downward with occasional brief recoveries, may be reached. Securities bought now will usually do well because they are made at historically low overall stock market levels. Another term, *bottoming out,* describes a process in which a long period of downtrend prices move in the opposite direction of an uptrend. The directional change can take place over varying time periods from several years to one day. The bottoming-out movement may be in a double or triple bottom, a rounding bottom, or other recognizable form.

FIGURE B.11—BULLISH AND BEARISH MARKET: SELLING CLIMAX

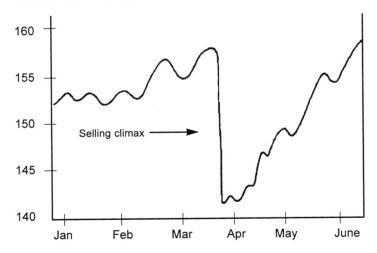

A bearish pattern may take the form of a descending triangle that is a sideways price pattern between two converging trendlines. The upper line is declining but the lower line is flat.

A selling climax is a substantial price decline coupled with extremely high volume resulting from panic selling of securities. A selling climax often occurs at the end of a bear market and is usually an optimal time to buy securities (see Figure B.11, page 45).

In a *bear trap* there is a false indication that an increasing trend in an overall index or particular stock has reversed.

BULLISHNESS INDICATORS

Surveys of investors seeking to determine whether market watchers and players are likely to be buyers (bulls) or sellers (bears) of stocks in the near term.

The Bridge Market Barometer is a sentiment indicator developed by canvassing a regular set of market participants to ask them if they are bullish or bearish on a given market one week from today. Sentiment readings typically have been used as a contrarian indicator, although in recent years many markets have shown that very high readings are actually leading bullish indicators.

Barron's publishes four such surveys each week: polls of investment advisers and traders by *Investors Intelligence*, Consensus Inc., and Market Vane, plus a poll of small investors by the American Association of Individual Investors. The Fidelity mutual funds of Boston, in conjunction with the University of Michigan, produces an investors' confidence index each month much like the university's well-known consumer confidence poll. The Fidelity index is covered in some daily newspapers. The Bridge Market Barometer is updated each Thursday and is available on http://www.bridge.com/front/.

Table B.2 shows how the Bridge Barometer fluctuated during the turbulent summer and fall of 1998.

Such polls can be used to measure the true strength of a market trend, often in a contrarian way. In a rising market or one near its peak, increasingly bearish feelings may indicate the market may be headed higher, since healthy skepticism abounds. A hot market with strong bullish feelings could on the other hand be a signal of overheated investor expectation; that might be a sell signal. In a falling market or one near a possible bottom, the slightest turn to bullishness could be a buying signal. At a low point for stocks, watch for any hint that bearish sentiment is ebbing.

TABLE B.2— HOW THE BRIDGE BAROMETER FLUCTUATES

Date	Bridge Barometer	% Bullish Response	Simple Figure Change
11/5/98	62.9	67%	+27
10/29/98	49.1	50%	-2
10/22/98	58.6	65%	+17
10/15/98	53.2	59%	+7
10/8/98	39.3	44%	-17
10/1/98	46.5	45%	-7
9/24/98	50.9	52%	+2
9/17/98	56.8	64%	+15
9/10/98	38.0	25%	-25
9/3/98	40.1	29%	-17
8/27/98	46.5	55%	-7
8/20/98	56.2	52%	+13
8/13/98	55.9	59%	+13
8/6/98	52.9	62%	+6
7/30/98	74.7	93%	+37
7/23/98	60.8	67%	+26
7/16/98	57.7	64%	+17
7/9/98	60.0	58%	+19
7/2/98	72.6	79%	+43

Source: *Bridge Information*

Such surveys often give conflicting signals, making analysis tricky. Do not lean towards the surveys of pro traders in such cases; many studies show that mom and pop investors often do quite well.

C

CANSLIM

Canslim is a method formulated by William O'Neill to select stocks for investment. He developed the following attributes of stocks that have done exceptionally well:

1. *(C)urrent quarterly EPS*. Stocks showing huge gains in appreciation had an increase in quarterly EPS of 20% or more relative to the same quarter in the previous year.

2. *(A)nnual earnings growth*. The 5-year EPS increased by an annual rate of 15% or more. A decrease in EPS for one year is acceptable as long as it is followed by a fast and significant recovery.

3. *(N)ew*. The stock price increases substantially because of something new, such as a product or service, a change in management, or a breakthrough in technology.

4. *(S)hares*: The best earning stocks had fewer than 25 million outstanding shares, the presumption being that the less the supply of shares, the higher the price. If there is not much of a float of shares, merely moderate buying may significantly move up price.

5. *(L)eader*. A leading stock in the industry usually performs best.

6. *(I)nstitutional ownership*. Institutional buying of a stock represents smart money. However, caution is in order. An extremely high percentage of institutional ownership of a stock (e.g., 80%) may result in a steep price decline if the institutions unload at once because of unfavorable news.

7. *(M)arket*. Most stocks do well when the overall stock market is doing well.

O'Neill's theory is "buy high and sell higher." He believes, based on his research, that the best time to buy a stock occurs when it moves into new high territory after a 2- to 15-month consolidation period. A sharp increase in the market price of the stock may take place after a breakout.

CATS AND DOGS

Very low-quality securities, usually having a low price.

CHANGES: UP AND DOWN

A listing, usually daily, of the stocks having the highest percentage increase in price and the highest percentage decrease in price for the trading period, in the order of their percentage change. Typically, 10 advancing stocks and 10 declining stocks are listed separately for the NYSE, AMEX, and NASDAQ. The listing appears in the financial pages of newspapers such as the Wall Street Journal.

CHANNEL

The area of the chart between the resistance and support levels (see Figure C.1). Some guidelines related to channels:

- Buy when a security is trading in a channel running significantly upward but is trading close to the low end of that channel, or when a security has been trading in a sideways channel and ultimately trades over its resistance level.
- Sell when a security is trading in a downward channel or its price goes below a support level.

Figure C.1 depicts a channel.

CHARTING

Charts are used to evaluate market conditions, including the price and volume behavior of the overall stock market and of individual securities. A daily chart is used to plot short-term minor moves. Daily charts over a one-year period are helpful in studying intermediate trends. A weekly chart is good to show volume and price ranges associated with previous major bull and bear cycles; intermediate tops and bottoms should form at the same ranges each year in bull and bear cycles.

FIGURE C.1—CHANNEL

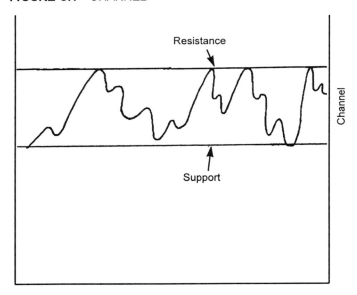

A false breakout is a breakout of a chart pattern aborting itself. The actual breakout should be substantiated by other technical indicators.

A cradle is the intersection of converting boundary lines of a symmetrical triangle. The axis support (or resistance) is strongest closest to the cradle point and becomes weaker as the axis line (apex level for resistance) extends out on the chart as time passes. If the axis support does not hold, the reaction may accelerate significantly in an "end run" around the line.

The three basic types of charts are line, bar, and point-and-figure. On line charts (Figure C.2) and bar charts (Figure C.3), the vertical line shows price and the horizontal line shows time.

On a line chart, ending prices are connected by straight lines. On a bar chart, vertical lines appear at each time period, and the top and bottom of each bar shows the high and low prices. A horizontal line across the bar marks the ending prices.

FIGURE C.2—LINE CHART

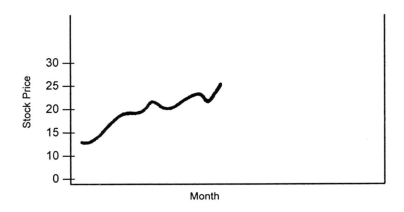

FIGURE C.3—BAR CHART

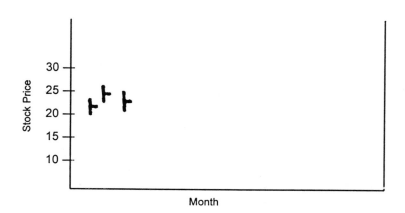

Point-and-figure charts (Figure C.4) show emerging price patterns in the market in general and for specific stocks. Typically, only the ending prices are charted. An increase in price is denoted by an X and a decrease by an 0.

FIGURE C.4—POINT-AND-FIGURE CHART

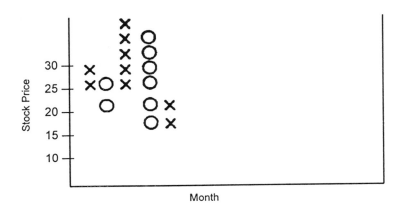

There is no time dimension in Figure C.4. A column of Xs shows an upward price trend while a column of Os reveals a downward price trend.

In point-and-figure charts, the price scale is vertical. Points are plotted on the chart when a price changes by a predetermined amount.

Significant price changes and their reversal are depicted. The individual investor decides what is significant. The investor can use either ending prices or inter-day prices, depending on time constraints. The usual set amounts are one or two points for medium-priced stocks, three or five points for high-priced stocks, and half a point for low-priced stocks. Most charts contain specific volume information.

The investor should plot prices representing a trend in a single column, moving to the next column only when the trend is reversed. He will usually round a price to the nearest dollar and start by plotting a beginning rounded price. Nothing new appears on the chart if the rounded price does not change. If the rounded price changes, the investor plots it. If new prices continue in the same direction, they will appear in the same column. A new column begins when there is a reversal.

Standard & Poor's Trendline gives charting information on many securities. Other financial services, financial magazines and newspapers, and brokerage research reports also provide charts. The Telescan Analyzer prepares graphs for technical and fundamental indicators.

The *Wall Street Journal* publishes a daily graph of the Dow Jones Industrials 30 along with many other charts of stock and market performance.

A chart pattern may be studied to predict future stock prices and volume activity. Charts may cover historical data of one year or less for active stocks or several years for non-active stocks.

Point-and-figure charts provide data about resistance levels (points). Breakouts from resistance levels indicate market direction. The longer the sideways movement before a break, the more the stock can increase in price.

The investor may use charts to analyze formations and spot buy and sell indicators. The investor can use these charts to determine whether the market is in a major upturn or downturn and whether the trend will reverse. The investor also can see what price may be reached by a given stock or market average. Further, charts can help the investor predict the magnitude of a price swing.

Caution: Historical trends in prices may not result in future price trends because of changing circumstances in the current environment.

COMMISSION

The fee assessed by a brokerage firm to buy or sell a security, commodity, or option. In the case of a commodity, the commission is assessed when the original entry trade is closed with an offsetting trade, referred to as a round turn commission.

COMMODITIES FUTURES

Contracts in which sellers promise to deliver a given commodity by a certain date at a predetermined price. The price is agreed to by open outcry on the floor of the commodity exchange. The contract specifies item, price, expiration date, and a standardized unit to be traded (e.g., 100,000 pounds). Commodity contracts may run up to one year.

Investors must continually evaluate the effects of market activity on the value of the contract. While the futures contract mandates that the buyer and seller exchange the commodity, the contract may be sold to another party before the settlement date when the trader wants to realize a profit or limit the loss. Investors engage in commodity trading in the hope of high returns and inflation hedges.

Commodities should be analyzed technically through the study of market fluctuation, mostly with the use of charts, to forecast future price trends. Usually, technical analysis focuses on the study of market action, while fundamental analysis concentrates on the supply and demand that make prices move higher, lower, or remain the same. Fundamentalists examine all factors that affect the pricing of a commodity to assign an intrinsic value to that commodity. The fundamentalist approach is to value a commodity based on the law of supply and demand.

The thrust of this analysis is to determine which way prices are likely to move. Usually, futures traders group themselves as either fundamentalists or technicians. Some principles of one usually apply to the other. Technicians normally have a limited amount of knowledge about the fundamentals of commodities trading. Fundamentalists usually have a decent grasp of the principles of chart analysis. There is another little-known group called purists who do not recognize fundamental data at all.

There are some conflicts between fundamentals and chartists in analyzing commodities data. The problem usually arises early in a critical market move, when there is no explanation for the actions the market is taking. An important time in the trend of the market, this is where the two schools of thought differ the most. After the market settles down, the two approaches seem to agree for a time—but usually too late for the trader to make an intelligent and logical decision.

The differences can be explained because market price is a leading indicator for the fundamentalists. Usually, the fundamentals are already included in the market price, because it reacts to missing fundamentals. Over time, these changes may occur so late that a new trend develops well after an important bull or bear market was started.

Technicians usually operate in a different manner; they rely heavily on their ability to use their charts, they have tremendous confidence in their charting ability. Technicians expect the normal thinking to be in opposition to the way the market is moving. They feel comfortable with this and know that the reasons the market is behaving this way will become known to everyone.

These two ways of thinking are drastically different. Most would opt for the technical approach, which appears the superior of the two. The fundamental approach is entwined in the technical approach, and to read a chart would include an abbreviated form of fundamental analysis. The trading of commodities can be accomplished by using only the technical approach without considering the fundamental side of the equation.

To trade commodities futures effectively, timing is important. Usually, the margin requirements on commodities are very low (normally less than 10%) so it is possible to understand the general trend of the market and still lose money. A small fluctuation in price in the opposite direction can drive the trader out of the market while losing either some or all of the trader's margin. To buy and hold is not the prudent approach to trading commodity futures.

Technical analysis can be adapted to any situation and applied to any type of trading and time frame. The technical analyst would be hard pressed to find a trading arena that these principles would not apply to. A technical analyst dealing with commodities can apply his or her principles to as many markets as needed, unlike the fundamentalist, who will concentrate on one market, such as frozen orange juice or gold, because of the tremendous amount of data that needs to be addressed,

Markets usually go through changes, with active and dormant periods, thus giving the technical analyst a direction to follow while disregarding non-trending markets.

Usually during the course of trading technical analysts get involved in the trending and active markets while disregarding inactive markets. Following these active periods are inactive periods where the technician will jump to another actively trading market. Fundamentalists usually concentrate on one market and do not exhibit the same kind of flexibility. Normally, the technician has a better overall picture of the market place than the true fundamentalist and realizes that price trends in one market may be the starting point for a trend in another.

Technical analysis also has uses in the wide-ranging field of economic forecasting. The Commodity Research Bureau (CRB) index has always been used as a leading indicator of inflationary trends. It has a close relationship with industrial production, commodity prices being spearhead movements in the index.

Analysts need to explore further the use of technical analysis as an indicator of sweeping economic trends.

A popular theory that relates to commodity futures is the Random Walk Theory, which is founded on the efficient market hypothesis. This means that prices move randomly around their intrinsic value, and that the best strategy would be to buy and hold as opposed to making a killing. The general feeling among analysts is that there is some degree of randomness but that there should be more reliance on statistical techniques and technical analysis.

There are some terms relating to commodity futures that the technical analyst should be familiar with. There are different types of trends. A *trend* is the movement of commodity prices in one direction. A *trend channel* is the likely parallel price range surrounding the probable price line. A trading range is depicted by horizontal peaks and troughs. Trends may be short-term, intermediate-term, or long-term. In a trending market, the price of a commodity is going in one direction, either upward or downward.

Bracketing is a range of commodity prices without a clear trend. *Uptrends* are a series of ascending peaks and troughs while downtrends are a group of descending peaks and troughs. *Running the market* occurs when commodity prices are in a fast continual upward or downward trend. There are also major and minor trends with the major trends extending over periods of two to three years. Minor trends usually reflect short-term variability in price, typically over a period not exceeding a few months. A *trading pattern* is a long-term direction in commodity prices, as shown in Figure C.5.

FIGURE C.5—TRADING PATTERN

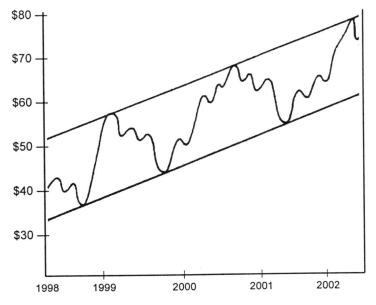

Support is a price area under the market where additional buying is expected (see the separate entry for Support and Resistance). *Resistance* is a price area above the market where additional selling is expected. Some markets will develop price channels that are parallel trendlines drawn under and over the price action (see Figure C.6).

FIGURE C.6—COMMODITY RESEARCH BUREAU FUTURES INDEX PRICE

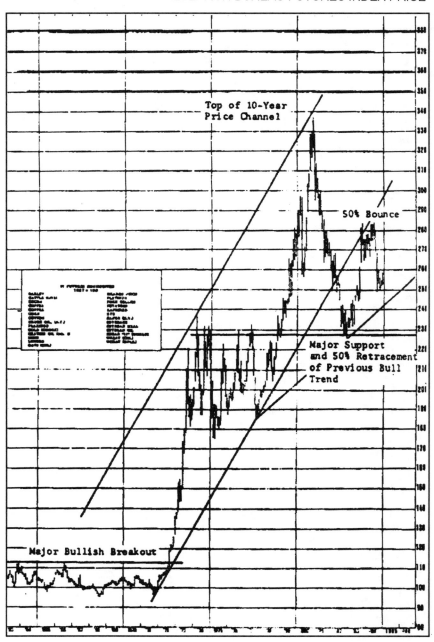

The wise technical analyst will use all the tools at his disposal in order to make wise trading decisions about commodity futures, putting together as many techniques as possible to seek clues to future market movement and price trends.

COMMODITY CHANNEL INDEX

Developed by Donald R. Lambert, the Commodity Channel Index (CCI) is a price momentum indicator. Despite the word commodity in the name, this indicator may be applied equally well to stocks and mutual funds. It measures price excursions from the mean price as a statistical variation. The CCI index formula is:

$$CCI = [M - MA]/.015D$$

$$\text{Where: } D = 1/n \left(\sum_{i-1}^{n} [Mi - MA] \right)$$

M = 1/3 (H + L + C) or the mean price for a period

H = highest price for a period

L = lowest price for a period

C = closing price for a period

MA = the n-period simple moving average of M

D = the mean deviation of the absolute value of the difference between the mean price and the simple moving average of mean prices.

The classic CCI registers a buy when it crosses the -100 line moving up, and a sell when it crosses the +100 line moving down. It was found that these buy and sell points sometimes missed trading opportunities. To correct for this, use the crossing of the zero line for buy and sell signals instead of -100 and +100. This is called the Zero CCI and this is what is shown in Figure C.7.

- When the Zero CCI crosses the zero line while moving up, a buy signal is generated.
- When the Zero CCI crosses the zero line while moving down, a sell signal is generated.

The CCI works best on strongly uptrending or downtrending stocks.

FIGURE C.7—THE COMMODITY CHANNEL INDEX (CCI) FOR MICROSOFT

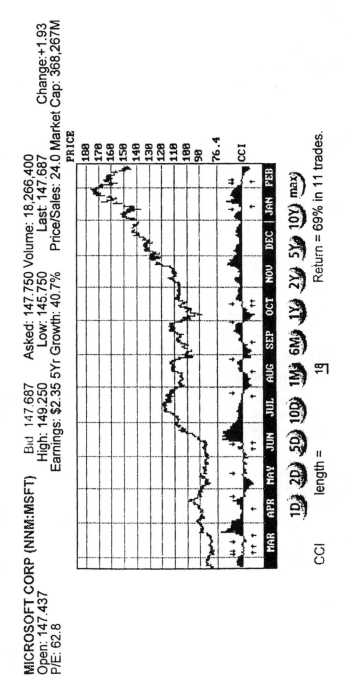

COMMODITY RESEARCH BUREAU INDEXES

The Commodity Research Bureau (CRB) has two indexes: the CRB Spot Price Index and the CRB Futures Price Index. The CRB Spot Price Index, based on prices of 23 different commodities, serves as an inflation indicator. The CRB Futures Price Index is a composite index (1967 = 100) that tracks the volatile behavior of 21 commodity prices. The CRB Futures Index, produced by Knight-Ridder Financial Publishing, is designed to monitor broad changes in the commodity markets.

Nine sub-indexes to the CRB Futures Index are maintained for baskets of commodities representing currencies, energy, interest rates, imported commodities, industrial commodities, grains, oil-seeds, livestock and meats, and precious metals.

The *Wall Street Journal* publishes monthly charts of the CRB Futures Index.

COMMODITY SOLD SHORT (SHORT SALES)

A promise to deliver the commodity at a specified price on a future date. Most commodity short sales are covered before the delivery date.

CONFIRMATION

1. The point where a security breaks outside a particular range.
2. A consistent move in two or more indices as evidence of a trend or conclusion. An example is when a new high or new low exists for both the DJIA and the Dow Transportation Average. A nonconfirmation is when only one index reaches a new high or new low. A nonconfirmation also occurs when there is an inconsistency in the movement of a stock market index and a technical indicator. An example is the Standard & Poor's 500 index reaching a new high coupled with fewer advances relative to declines over the previous two weeks. Failure of major technical indicators to confirm the new high in the S&P index is a bearish sign.
3. Two or more technical indicators reaching the same conclusion. An example is when volume and price move together.

CONGESTION AREA

A period of sideways (horizontal) price movement of the overall stock market on a chart within the bottom and the top. By appraising the congestion area, the investor may be able to ascertain the future direction of the breakout. A *fulcrum* is a congestion area occurring after a substantial advance or decline that forms an accumulation base or a distribution top. A *catapult* is a breakout over the top of a congestion area.

CONSOLIDATION

A period when there is a balanced relationship between demand and supply for securities (stocks, bonds) in which prices move in a narrow range. If the relationship becomes unbalanced, prices will move outside the range. Examples of consolidation patterns might be rectangles and triangles.

CONTINUATION FORMATION

A pattern in which demand and supply for a stock is relatively equal. This formation occurs if the stock price approaches from one direction and exits in the same direction.

CONTRARIAN INVESTING

CONTRARY OPINION RULE

Contrary opinion is a sentiment indicator in which after finding out what most investors are doing, the investor does the opposite. The rationale is that popular opinion is usually wrong. The rule presumes that the crowd is typically incorrect at major market turning points.

The majority opinion is normally reflected in the media, including *Business Week, Forbes, Fortune, Newsweek,* and *Time,* financial books for lay people, and television financial news programs.

Published and televised news usually has been reflected in the price of securities. Therefore, publicizing of the news is probably the end instead of the beginning of a move.

If most investors are bullish, they have probably fully invested their available funds—contributing to a stock market peak. If most investors are bearish, they have probably sold their shares—contributing to a stock market bottom.

If everyone is pessimistic, the contrarian investor concludes that it is probably the time to buy. If everyone is optimistic, the investor believes that it is probably the time to sell.

Investors should compare the news stories with other technical and fundamental indicators. They may find good buys for company stocks that are out of favor because of an oversold situation. However, these stocks should possess fundamental values based on the company's financial condition.

Caution: There may be instances in which what most investors are doing is the right strategy. For example, the investing public may be buying securities during what is in fact a bull market.

INDEX OF BEARISH SENTIMENT

This index is based on a reversal of the recommendations of investment advisory services as contained in their market letters. Such services are considered to be a proxy for majority opinion. *Investor's Intelligence* believes that advisory services are trend followers rather than anticipators, recommending equities at market bottoms and offer selling advice at market tops. This index operates according to the contrary opinion rule: Whatever the investment advisory services recommends, the investor should do the opposite. It is a technical investment analysis tool.

The index is computed as follows:

$$\text{Index} = \frac{\text{Bearish investment advisory services}}{\text{Total investment advisory services}}$$

Investors Intelligence believes that when 42% or more of the advisory services are bearish, the market will go up. On the other hand, when 17% or fewer of the services are bearish, the market will go down. The Index of Bearish Sentiment, originally developed by A.W. Cohen of Chartcraft, is published by *Investors Intelligence* and can be found in *Barron's*.

Example: Of 200 investment advisory services, 90 are bearish on the stock market. The Index equals 0.45 (90/200). Since 45% of the advisory services are pessimistic about the prospects for stock, more than the 42% benchmark, the investor should buy securities.

A movement toward 10% means that the Dow Jones Industrial Average is about to go from bullish to bearish. When the index approaches 60%, the Dow Jones Industrial Average is headed from bearish to bullish.

If sentiment is bearish, a bull market is expected, and the investor should buy stock. If sentiment is bullish, a bear market is likely, and the investor may consider selling securities owned.

Caution: Other measures of stock performance should be used in conjunction with this index.

CORRECTION

A short-term price move in a stock, bond, commodity, or index opposite that of the prevailing trend but not of a magnitude to change the long-term trend. In a downtrend it is referred to as a rally, in an uptrend as a reaction. Figure C.8 depicts a correction.

FIGURE C.8—CORRECTION

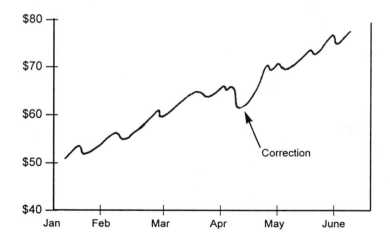

CYCLE ANALYSIS

A cycle is a price pattern of movement, event, or phenomena that regularly takes place over a specified time period. However, changing expectations over time influence a pattern. The greater the length of time an indicator is at a particular value, the more predominant (stronger) it is. Cycle analysis improves the predictability of future developments in stock, bond, and commodity markets. However, cycles are complicated to evaluate and can be distorted by random occurrences; they may be off course at times.

Cyclical investing is the purchase or sale of securities based on the point in the long-term market cycle. A cyclical stock tends to increase quickly when the economy improves and falls quickly when the economy deteriorates. Examples are paper, automobiles, and housing. Non-cyclical stocks are not as directly impacted by economic changes. Examples are foods and pharmaceuticals.

Short-term daily chart cycles have a length of less than 30 days. The near-term trend structure is influenced by the direction of the long-term and intermediate trends, especially at times of correction or recovery against the underlying dominant trend.

Figure C.9 shows a chart illustrating a cycle.

FIGURE C.9—CYCLE ANALYSIS

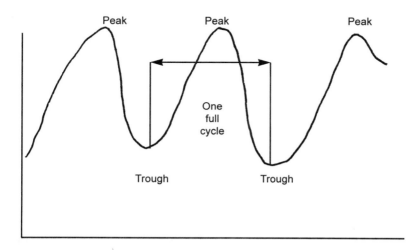

Some believe that there is a nine-year cycle associated with the economy. Others have different choices, e.g., 40 months, 4 years, 8 to 10 years. Whatever the length of the cycle, a cyclical industry sees its earnings go up and down based on the economic cycle. Companies in a cyclical industry do well when the economy is prospering, poorly when it is deteriorating.

The business cycle is the regular pattern of expansion (recovery) and contraction (recession) in aggregate economic activity around the path of trend growth, with effects on employment and inflation. At the peak of the cycle, economic activity is high relative to trend, whereas at the trough (valley) of the cycle, economic activity reaches its low point. The business cycle tends to have an impact on corporate earnings, cash flow, dividends, and expansion. One cycle extends from a gross national product (GNP) base line through one rise and one decline and back to the base line, a period of about two and a half years.

The business cycle approach to timing stock selection relates to factors external to the stock market itself, such as industrial production and the availability of money in the economy. The technician attempts to improve timing decisions by examining market factors such as price and trading volume. Here, technical analysis applies to internal analysis or market analysis.

Think about the life cycle of a given industry. What point in the life cycle is the industry currently in? It is best to get in at the beginning when there is very profitable expansion. This stage is often followed by slow growth, maturity, and decline. Stock price will rise, remain static, and then fall during the course of the life cycle.

The January cycle (effect) suggests that there is a tendency for the stock market to be higher at year-end if stock prices increase during January, and lower if they decrease. In other words, look at January to determine how stocks will fare for the year.

In *the Presidential cycle* security prices decrease immediately after a new president is elected because he will most likely take unpopular economic actions. However, during the mid-term period, stock prices start to increase because of an expected stronger economy due to the initial economic steps.

Some believe there is a 40- to 53-month cycle associated with the stock market. The up-wave has a stronger economy and a mild bull market. When the plateau is reached, the economy has peaked with a strong bull market. The down-wave then starts, leading to a bear market.

Specific commodities experience unique seasonal cycles, notably the cycle of 9-12 months exhibited by the Commodity Research Bureau

Index. Some believe in a 28-calendar day cycle applied to the wheat market. Other examples of cycles in the form of data sets of periodicity are:

- 17½-year cycle in U.S. wholesale prices
- 20-month cycle in soybean prices
- Sugar cycle
- Shoe manufacturing cycle
- 18-year cycle in coal production
- Industrial bond yield cycle
- 6-year cotton cycle
- Copper price cycle
- Pig iron cycle
- Automobile factory sales cycle
- Fashion cycle
- Weather cycle
- 45-year drought cycle

J.M. Hurst characterizes the stock market as wave rhythms acting together in cycles of nominal length. The wave averages are helpful in formulating a predictive model signaling the likelihood of cyclic lows.

There is a relationship between cycles. For example, two 10-week nominal cycles make up one 20-week normal length. There is a tendency for cyclic rhythms to nest at major market bottoms.

Fast Fourier Transformation (FFT) deals with the amplitude and length of cycles. It can derive the major cycles from a data series such as stock prices. The method "detrends" data based on either a moving average or a linear regression. The method also smoothes prices. A software program may be used to extract the major cycle lengths (in trading days) along with their relative strength.

Figure C.10 presents a chart illustrating Fast Fourier Transformation. It reveals that the most significant (left) cycle is 110 days, while the least significant (right) cycle is 15 days.

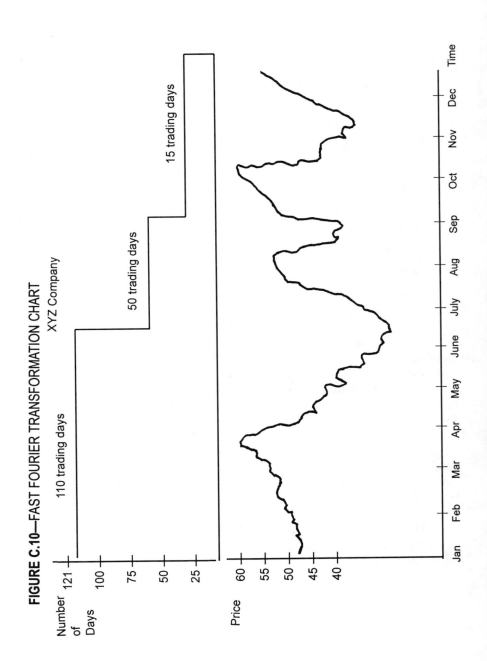

FIGURE C.10—FAST FOURIER TRANSFORMATION CHART

XYZ Company

D

DEMAND *VERSUS* SUPPLY

The majority of technical analysis approaches are based on the demand/supply relationship.

Demand is the total number of shares buyers want to purchase, and amount that continually changes. If demand equals supply, stock prices should remain constant. If the amount demanded significantly exceeds supply, stock prices will increase. Accumulation may occur.

Supply is the total number of shares available for sale, an amount that varies over time. When supply significantly exceeds demand, stock prices will decline. Distribution may occur.

A *soft market* is characterized by the excess of supply over demand. A soft market in stock is defined by inactive trading, significant spread between bid and offer, and sharp price drops in response to minimal selling pressure.

Just a moderate oversupply (extra demand) at a particular price will be sufficient to stifle (stem) an advance (decline). *Overhanging supply* refers to stock purchased at higher levels by holders now waiting for an opportunity to unload their shares. *Floating supply* refers to the number of outstanding shares that are not closely held by companies or individuals who wish to retain their ownership in the entity. Floating supply therefore equals outstanding shares minus closely held shares and shares not available to the public (e.g., shares held by an employee pension fund).

Demand/supply analysis is concerned much more with short-term disparities than with long-term trends. Demand/supply analysis also attempts to measure imbalances between new stock offerings and expected investment demand for common stocks. An excess of offerings compared to demand will depress stock prices.

A stock chart helps in gauging the relative strength of demand and supply of a particular stock or other market instrument. The demand/supply line presents what the demand/supply relationship will be at a specified price. The demand line represents the number of shares buyers are willing to buy at specified prices. If prices decrease, buyers will be willing to buy more shares at the lower prices. The supply line depicts the number of shares sellers are willing to supply at a particular price. If prices are going up, sellers would be more inclined to sell.

In an *equilibrium market*, the prices of securities in the stock or commodity market reflect a balance between demand and supply. There is an equilibrium price (see Figure D.1) when the supply price of a stock matches the demand price.

FIGURE D.1—DEMAND VERSUS SUPPLY

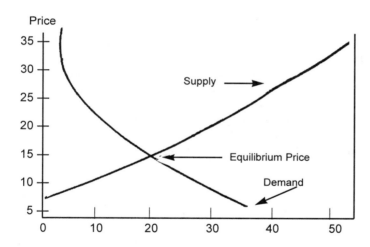

A supply/demand chart indicates how many buyers and sellers there are at a particular price. In Figure D.2, when the price is $35, there are 15 buyers and 25 sellers.

Figure D.2 shows support is where the supply line intersects the vertical axis, at a price of $24. Prices will not go below $25, since no sellers want to sell at less than $24. There is resistance where the demand line touches at the $47 price on the vertical axis. Prices cannot go above $47 because there are then no willing buyers.

FIGURE D.2—DEMAND/SUPPLY

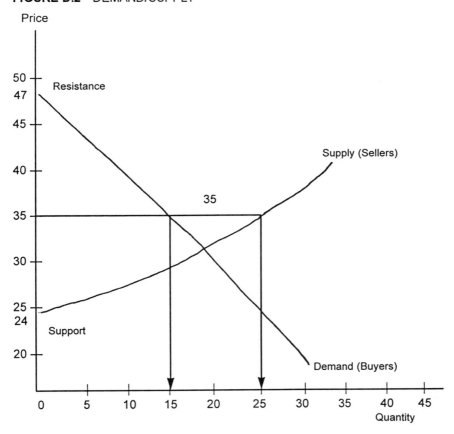

In a free market, the lines and prices will change over time based on changes in investor expectations. If the support level breaks, it means the supply line will go downward because more sellers are willing to sell at lower prices. The converse occurs when the resistance level breaks: The demand line will go up because more are willing to buy at higher prices.

Area patterns or formations of price variability along with related volume indicate key changes in the balance of demand and supply. They may reveal consolidation, recuperation, or gathering of strength for movement in a similar direction as the previous trend. On the other hand, they may depict a reversal in an opposite direction.

A congestion area is the sideways (flat, horizontal) trading range in which demand and supply balance out. This is different from the consolidation area in which there is a trading range in which prices move until continuing in the same direction as the trend before the consolidation area.

James Sibbet has developed a *demand* index that takes into account the interaction of volume and price to predict future change in prices. A divergence between prices and the demand index is an indicator of forthcoming price weakness. A significant divergence between the demand index and prices points to a major bottom or top.

Since the demand index is a leading predictor, a high demand index may indicate that prices will rally to new highs. The index may also act as a coincidental indicator in that a lower demand index peak coupled with higher stock prices may indicate a top. The index may also serve as a lagging indicator because when the demand index penetrates the zero level a change in trend is signaled. If the demand index is close to the zero level for a while, it implies that weakness in prices will not last long.

DESCENDING TOPS

See Ascending and Descending Tops.

DIP

A minor decrease in stock prices after a sustained uptrend. Some investors buy on dips when the stock price is temporarily weak (see Figure D.3 on page 73).

DISTRIBUTION

See Accumulation and Distribution.

DIVERGENCE

A divergence occurs when the price of an item or measure (e.g., stock, bond, commodity contract, index) moves significantly in a direction that is either not confirmed or not accompanied by a similar move in a related market indicator. The size of the divergence is paramount.

FIGURE D.3—DIP

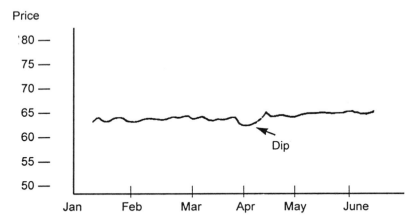

An example of divergence for the Dow Theory might be when there is a new high in the Dow Jones Industrial Average (DJIA) but not in the Transportation Average. In this case, there is a negative divergence, a bearish sign. *Negative divergence* occurs when two or more averages, indicators, or indices do not depict confirming trends. *Positive divergence* occurs when the Industrial Average is at a new low while the Transportation Average is not. This is a bullish sign. Many indices may be appraised for divergence, particularly momentum indicators relative to the Dow Jones Industrial Average or Standard and Poor's (S&P) 500 Index. Obviously, the more measures that confirm a divergence, the greater the significance.

Negative divergence may exist at a market top, while at a market bottom there may be positive divergence. In other words, one stock market index (DJIA) may show a new high or low that is not realized by another comparable measure (e.g., S&P 100 Index). Divergence often indicates a trend reversal. In oscillator appraisal, prices may move lower while an oscillator begins to increase.

Divergence exists if the trend in the price of a security does not match the trend in an indictor. When there is divergence, prices typically change direction to confirm the trend of the indicator, because as a measure of price trends, the latter is superior to the prices themselves.

DOLLAR-COST AVERAGING

Dollar-cost averaging may be used for stock considered to be a good long-term investment. A constant dollar amount of stock is bought at regularly spaced intervals. The strategy represents time diversification. By investing a fixed amount each time, more shares are bought at a low price and less shares are bought at a high price. The usual result is a lower average cost per share because the investor buys more shares of stock with the same dollars.

The technique is advantageous when a stock price moves within a narrow range. If there is a decrease in price, the investor will incur less of a loss than ordinarily. If there is an increase, the investor will gain, but not as much as usual. Drawbacks to the method are:

1. Dollar-cost averaging has higher transaction costs; and

2. It will not work when stock prices moving are in a continuous downward direction.

Dollar-cost averaging is a conservative investment strategy since it avoids whimsical behavior, as when the investor is tempted to buy when the market is high or sell when the market is low. A conservative stock may be acquired with relatively little risk benefiting from long-term appreciation. Further, the investor is not stuck with too many shares at high prices. In addition, in a bear market many shares can be bought at very depressed prices.

Example: Every month an investor invests $100,000 in ABC Company, making the following transactions:

Date	Investment	Market Price	Shares Purchased
6/1	$100,000	$40	2,500
7/1	100,000	35	2,857
8/1	100,000	34	2,941
9/1	100,000	38	2,632
10/1	100,000	50	2,000

The investor has purchased fewer shares at the higher price and more shares at the lower price. With the $500,000 investment, 12,930 shares have been bought, resulting in a average cost per share of $38.67. On October 1, the $50 market price of the stock thus reflects an attractive gain.

Investors seeking to use this strategy must determine how much they can afford to contribute systematically to a dollar-cost averaging account each month or quarter.

DOUBLE BOTTOM OR TOP

A *double bottom* is a chart pattern depicting a decrease in price, then a rebound, than another drop to the same level. The pattern is typically interpreted to mean the stock must have a lot of support at that price and should not fall further. If, however, the price goes below that level, it is deemed a new low. Figure D.4 presents a double bottom.

FIGURE D.4—DOUBLE BOTTOM

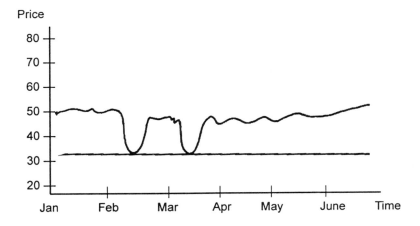

A *double top* is a chart pattern reflecting a rise to a high price, then a drop, then another rise to the same high price. This means the stock has experienced resistance to a move higher. If, however, the price goes through that level, the security is likely to reach a new high. Figure D.5 presents a double top.

FIGURE D.5—DOUBLE TOP

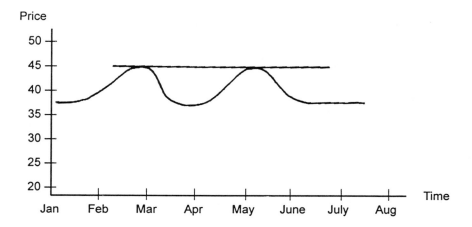

DOUBLE CROSSOVER

A double crossover occurs when a buy signal arises because the shorter average crosses above the longer one.

DOW THEORY

The Dow Theory, in existence since the late 1800s, was developed by Charles H. Dow in a series of articles published in the *Wall Street Journal*. (Charles H. Dow founded the Dow-Jones Financial News Service, which is credited with inventing stock market averages.) Early in the 1900s, William H. Hamilton, Dow's replacement as editor of the *Wall Street Journal*, refined his predecessor's principles into what is known as the Dow Theory (see Figure D.6).

The Dow Theory states that most stocks move in the same direction as the overall market. If the stock market is on an upswing, most stocks will move upward; if the market is on a downward swing, most stocks will be moving downward. An uptrend experiences a pattern of rising peaks and troughs, a downtrend a pattern of falling peaks and troughs.

The Dow Theory helps investors determine whether a major trend in the market is occurring and in what direction. The theory originally was concerned with general stock market trends as a measurement for general business conditions. It was not developed to forecast stock prices, but current use is focused almost exclusively for forecasting.

FIGURE D.6—DOW THEORY CHART

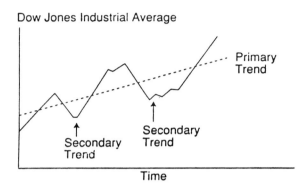

To measure the overall market Charles Dow developed two market indexes, now known as the Dow Jones Industrial Average and the Dow Jones Transportation Average.

The Dow Jones Industrial Average includes 30 major industrial corporations such as IBM, GE, and GM. The Dow Jones Transportation Average includes 20 major transportation corporations such as Consolidated Freightways, Union Pacific Railroad, and Federal Express. Originally, in the late 1800s Dow used only railroad companies because the railroads were the thriving businesses of that era; the industrial average was made up of the stocks of 12 businesses of all other types. Over the years the stocks that have comprised these lists have changed to keep the list representative of the current market. The only stock in the current Dow Jones Industrial Average that has been included from inception is General Electric.

In 1929 utility stocks were separated from the Industrial Average and included in an average of their own, the Utility Average. At present, the 15 utility stocks, 30 industrial stocks and 20 transportation stocks are also combined and averaged to produce the Dow-Jones 65-Stock Composite. The Dow theory only includes the industrials and transportation stocks, not the utilities or the composite average.

Today the Dow Jones averages are calculated hourly as well as daily as published in financial papers such as the *Wall Street Journal.*

Figure D.7 on page 79 shows a comparison of the two senior Dow Jones Averages for a two-year period. They usually move in the same direction.

The Dow Theory may be summarized as follows:

1. It is based on closing prices.

2. A trend stays in effect until a reversal is indicated by both averages.

3. A bull trend is indicated when both averages go above their highs of previous upward secondary reactions. A bear trend exists if both averages fall below their lows of previous secondary reactions. In most cases, one average indicates a change in trend before the other. Under the Dow Theory, there is no specified time to invalidate a confirmation.

4. Demand and supply factors for stocks have been discounted by the averages.

5. Primary downtrends reflect a bear market. There are typically three down moves. The first is when investors believe corporate profits cannot be maintained so they sell stocks. In the second, panic sets in with sellers quickly exiting the market and few buyers coming in. In the third, financially-strapped sellers are forced to raise cash.

6. Primary uptrends reflect a bull market. There are typically three up moves. In the first, investors buy stock when business conditions are expected to get better. In the second, investors buy stock because of higher corporate profitability. In the third, there is much speculation.

7. Volume should expand in the direction of a major trend. This means that volume must confirm the trend.

8. A trend is presumed to be in effect until it provides definite signals of reversal.

FIGURE D.7—DOW JONES INDUSTRIAL AVERAGE (TOP) AND DOW JONES
TRANSPORTATION AVERAGE (BOTTOM)

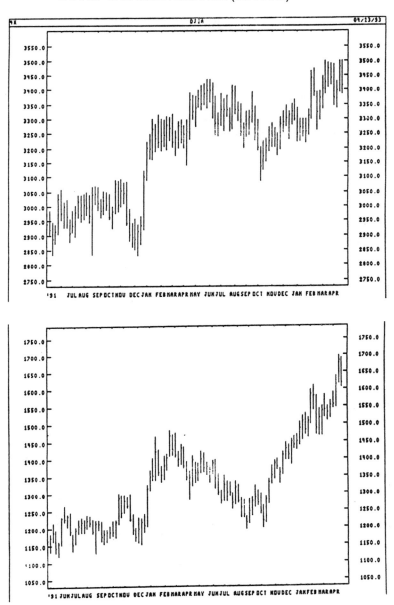

The Dow Theory is based on several important principles:

- *The averages discount everything.* The market discounts everything including economic and psychological factors. The Dow averages indicate the combined market activities of many investors, including many knowledgeable ones with excellent foresight and current information on events and stock market trends. Even natural disasters are quickly dealt with and their effects discounted.

- *There are three trends in the market.* The price of stocks moves in trends, the most important being major or primary trends that represent the long-term trend in the price of stocks. These trends are broad up or down movements lasting a year or more and culminating in a general appreciation or depreciation in value of greater than 20%. A primary or major trend is more commonly known as a bull (rising) or a bear (falling) market. A bull market is a wide-sweeping upward movement usually averaging about 18 months and broken by secondary reactions.

When the worst possible news has been taken into account and the outlook for the future begins to be positive, a bull market has started. The prices of equities usually increase with improvements in business conditions until finally stock prices are advanced due to unfounded projections.

A long decline stopped by significant rallies is a bear market. It starts on the abandonment of the first stocks to be purchased, moving towards declining levels of profits and business activities. The bear market ends when investors sell off holdings regardless of their value.

Secondary reactions move in the opposite direction, interrupt the primary trend, and tend to be a corrective force. These occur when a primary move has exceeded itself. Secondary trends are comprised of minor trends or fluctuations that occur day by day. Secondary trends usually retrace the preceding primary price change from 33%-66%. This relationship is shown in Figures D.8 and D.9.

There are also minor trends that last from a few hours to as long as three weeks. They form sections of the primary or secondary moves and have virtually no forecasting value for the long-term investor. This is the only one of the three trends that can be manipulated.

- *Lines indicate movement.* A line is a price movement two or three weeks or longer during which time the prices of both Dow Averages fluctuates within a range of about 5%. This move shows accumulation or distribution of holdings.

FIGURES D.8 AND D.9—SCHEMATIC DIAGRAMS OF DOW THEORY BEAR MARKET SIGNALS

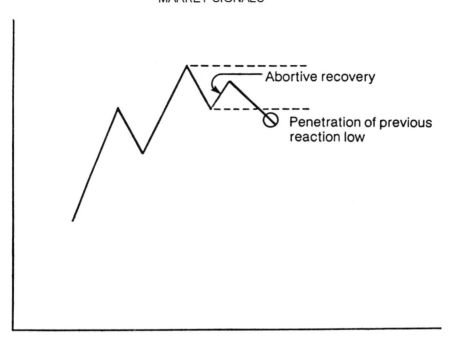

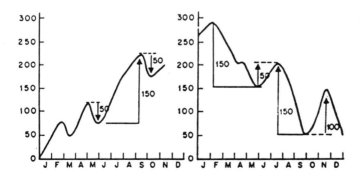

- *Price and volume are related.* Normally, volume expands on rallies and contracts on declines. If the opposite occurs—if volume decreases as prices increase or volume decreases on a price decline—a red flag will be raised, indicating that the present trend will soon be reversing.

- *Price fluctuation determines the trend.* Indication of a bullish trend is noticeable when repetitive rallies infiltrate peaks while the trough of a current decline is eclipsing the preceding trough. On the other hand, a number of falling peaks and troughs would indicate a bear market. The chance of a significant reversal will be higher if the market has experienced false hopes and speculation while the market is in an upswing. The opposite can also be true— the chance of an important reversal is greater if there is rampant pessimism and extensive liquidation during a period of market decline.

- *The averages must confirm each other.* Probably the most significant principle of the Dow Theory is that the trends of the Industrial Average and of the Transportation Average must be moving together in the same fashion (see Figure D.10). This confirmation makes sense in that investors should be driving up the prices of companies that manufacture goods and also of companies that transport those goods. To manufacture goods and not transport them to a salable location would not make sense in an expanding economy.

FIGURE D.10—EXAMPLES OF INDUSTRIAL AND TRANSPORTATION AVERAGE CONFIRMATIONS

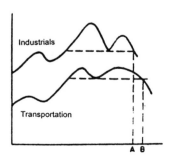
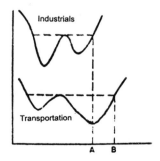

A hypothetical chart of a confirmation of the two averages is presented in Figure D.11.

FIGURE D.11—A HYPOTHETICAL CHART OF A CONFIRMATION OF
                        THE TWO AVERAGES

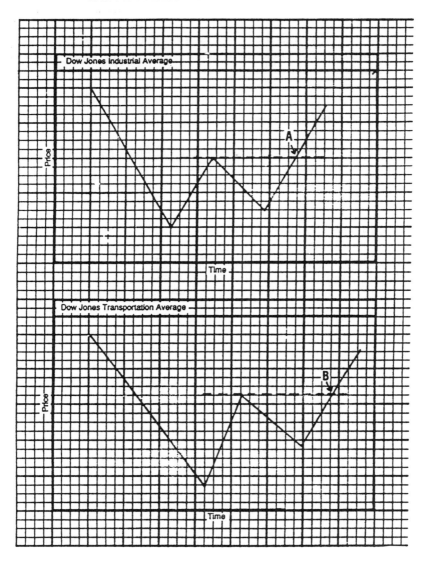

The Dow Theory has its share of critics as well as supporters. Some critics argue that the Dow Theory is too vague because it fails to state what the length and magnitude of the primary, secondary, and minor trends should be. This leads to varied interpretations of the direction of the primary trend due to differences in how it is viewed. Some consider the Dow Theory a late indicator of the beginning or end of a major market move.

Another problem with the Dow Theory is that it fails to reveal short-term trend signals, so short-term investors are not among its beneficiaries.

In summary, the Dow Theory implements the following strategies:

1. Buy when the market moves higher than the last peak.

2. Sell when the market moves below the preceding valley.

Generally, the Dow Theory has been a help rather than a hindrance. Sometimes it gives inaccurate signals, but overall it has been a friend to investors.

DOWNTURN

A shift in the stock market cycle from increasing to decreasing. The stock market is in a downturn when it goes from a bull to a bear market. The economy is in a downturn when it goes from expansion to recession.

EASE OF MOVEMENT

The comparison of volume to price can be examined by the ease of movement (EMV) indicator. The indicator reveals how much volume is needed to move prices. If prices are increasing on light volume, there is high ease of movement; if they are decreasing, ease of movement is low. The indicator will be closer to zero when prices are not moving or significant volume is needed for price movement. If the indicator goes above zero, that is a buy sign because prices are increasing more easily; an indicator below zero is conversely a sell sign. A smoothing with a moving average is usually done for ease of movement. The following steps are involved in computing ease of movement:

1. Determine the midpoint move as follows:

$$\frac{\text{Today's high + Today's low}}{2} - \frac{\text{Previous days high - Previous day's low}}{2}$$

2. The box ratio equals:

$$\frac{\text{Volume (in 10,000s)}}{\text{High - Low}}$$

3. Ease of movement equals:

$$\frac{\text{Midpoint move}}{\text{Box Ratio}}$$

EFFICIENT MARKET HYPOTHESIS

The efficient market theory presumes that the market is efficient and discounts all influences on it. Because of many varied, diverse, and shifting influences on the market, it is not feasible to detect a discernible pattern. Stock price is assumed to reflect immediately and fully everything known about the stock, new developments, and any expectations. The market price of a stock is identical to its real (intrinsic) value. A price change is equally likely to be positive or negative. The investor cannot regularly outperform the stock market because securities are presumed to be properly priced. In other words, the investor cannot beat the market. Because of the random nature in which information arrives, technical or fundamental analysis yields little. Past information has no impact on future prices. Under the efficient market hypothesis, the best investment strategy is buy and hold.

An efficient market may be weak, semi-strong, or strong.

In the weak form, there is no relationship between past and future stock prices. The value of historical information already lies in the current price. Therefore, there is no reason to review past prices.

This puts into question the very nature of technical analysis.

In the semi-strong type, stock prices immediately reflect new information. Therefore, action after a known event produces random results. All public information is incorporated into a stock's value. Therefore, fundamental analysis is not helpful in ascertaining whether a stock is over- or undervalued. Investors quickly consider information.

The strong form suggests that stock prices reflect all information, whether it be public or private (insider). A perfect market exists. There is no individual who has sole access to information. Thus, superior returns cannot be earned by one individual or group of individuals.

The hypothesis applies most directly to large companies trading on the New York and American Stock Exchanges.

Tip: Investors who uncover inefficiency in the market such as in stocks of small, little-known businesses with significant growth potential can profit by exploiting that information.

EFFICIENT PORTFOLIO

A portfolio that has a maximum expected return rate for any risk level. Its derivation considers the expected return and standard deviation of returns for each security, and the covariance of returns between different stocks in the portfolio.

ELLIOTT WAVE ANALYSIS

Named after Ralph Nelson Elliott, Elliott Wave Analysis is an approach
to market analysis based on repetitive wave patterns and the Fibonacci
number sequence. Inspired by the Dow Theory and by observations found
throughout nature, Elliott concluded that the movement of the stock market could be predicted by observing and identifying a repetitive pattern of
waves. In fact, Elliott believed that all human activities, not just the stock
market, were influenced by these identifiable series of waves. With the
help of C. J. Collins, Elliott's ideas received the attention of Wall Street
in a series of articles published in *Financial World* magazine in 1939.

The Fibonacci number sequence (1, 2, 3, 5, 8, 13, 21, 34, 55, 89,144,
233, etc.; 0 + 1 = 1, 1 + 1 = 2, 2 + 1 = 3, 3 + 2 = 5, 5 + 3 = 8, 8 + 5 = 13,
etc.) is constructed by adding the first two num-bers to arrive at the third.
The ratio of any number to the next larger number is 62 percent, which is
a popular Fibonacci retracement number. The inverse of 62 percent,
which is 38 percent, is also used as a Fibonacci retracement number. The
ratio of any number to the next smaller number is 1.62 percent, which is
used to arrive at Fibonacci price targets.

The connection between Elliot's contention of repeating cycles of
natures and the Fibonacci number sequence is that the Fibonacci numbers
and proportions are found in many manifestations of nature. For example,
a sunflower has 89 curves, of which 55 wind in one direction and 34 in
the opposite direction. Elliot observed in an 80-year period that the market showed a five-wave advance followed by a three-wave decline. He
concluded that a single cycle comprised eight waves (five up and three
down), as shown in Figure E.1 (3, 5, and 8 are Fibonacci numbers). The
eight waves are here labeled 1, 2, 3, 4, 5, a, b, and c.

More specifically:

1. Action is followed by reaction.

2. There are five waves in the direction of the main trend followed by
   three corrective waves (a "5-3" move).

3. A 5-3 move completes a cycle. This 5-3 move then becomes two
   subdivisions of the next higher 5-3 wave.

4. The underlying 5-3 pattern remains constant, though the time span
   of each may vary.

FIGURE E.1—FIBONACCI CYCLE

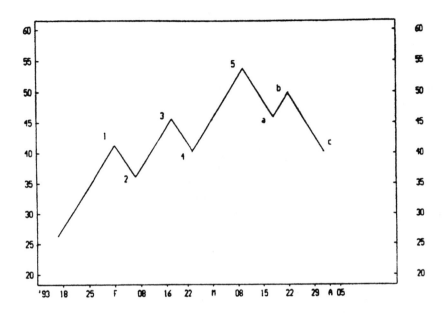

Waves 1, 3, and 5 are called impulse waves. Waves 2 and 4 are called corrective waves. Waves a, b, and c correct the main trend made by waves 1 through 5. The main trend is established by waves 1 through 5 and can be either up or down. Waves a, b, and c always move in the opposite direction to waves 1 through 5.

The theory holds that each wave within a wave count contains a complete 5-3 wave count of a smaller cycle. The longest cycle in the Elliot concept is called the Grand Supercycle. Each Grand Supercycle wave can be subdivided into eight cycle waves. The process continues to embrace Primary, Intermediate, Minute, Minuette, and Sub-minute waves.

Figure E.1 shows how 5-3 waves are comprised of smaller cycles. This chart contains the identical pattern shown in the preceding chart (now displayed using dotted lines), but the smaller cycles are also displayed. For example, you can see that impulse wave labeled "1" in the chart is comprised of five smaller waves.

Fibonacci numbers provide the mathematical foundation for the Elliott Wave Theory. Each of the cycles that Elliott defined consists of a total wave count that falls within the Fibonacci number sequence. For example, the chart shows two primary waves (an impulse wave and a corrective wave), eight intermediate waves (the 5-3 sequence shown in Figure E-1), and 34 minute waves (as labeled). The numbers 2, 8, and 34 fall within the Fibonacci numbering sequence.

Elliott Wave practitioners use their determination of the wave count in combination with Fibonacci numbers to predict the time span and magnitude of future market moves ranging from minutes and hours to years and decades.

Caution: The Elliott Wave Theory's predictive value is dependent on an accurate wave count. Determining where one wave starts and another wave ends can be extremely subjective.

EXHAUSTION

The condition in which buying power is insufficient to move prices higher, or selling power is insufficient to move prices lower.

FINANCIAL PYRAMID

A financial pyramid (see Figure F.1) is a structure of low-, medium-, and high-risk investments. In a financial pyramid the most money is invested in liquid and safe investments while the least money is in high-risk vehicles. As risk increases so does the rate of return.

FIGURE F.1—FINANCIAL PYRAMID

FIVE-DAY-AWAY RULE

An arbitrary time frame identifying a minor top or bottom in the market.

FLAG

A technical chart pattern resembling a flag shape with masts on either side, showing consolidation within a trend (see Figure F.2). A flag arises from price variability within a narrow range. It is preceded and followed by sharp rises or declines. If the flag (consolidation period) is preceded by a rise, it will typically be followed by a rise; a fall will usually follow a fall. A flag is a continuation price pattern lasting usually for no more than three weeks. It looks like a parallelogram sloping against the prevailing trend. A flag is a minor pause in a dynamic price trend.

FIGURE F.2—FLAG

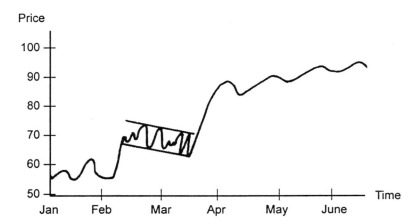

Flags and pennants that last too long (e.g., five weeks) are questionable. Stops should be set at the typical difference above or below the extreme tops or bottoms. Hold a stock if the move leading to the flag is up; if the move is down, retain the short position.

A *pole* is a sharp, practically vertical descent or ascent in stock prices typically occurring before a flag or pennant pattern.

FLAT MARKET

A flat market is characterized by horizontal price movement (see Figure F.3) It typically arises from low activity. In a flat market, prices trade within a narrow range over an extended period, showing only small up or down changes. An example is a stock trading between $40 and $42 for a year.

FIGURE F.3—HORIZONTAL PRICE MOVEMENT

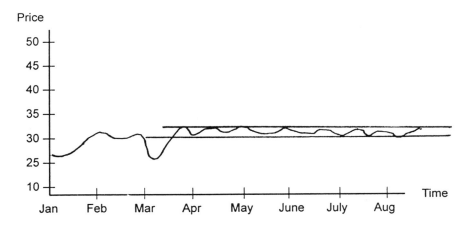

FOREIGN CURRENCY TRADING

The foreign exchange market is a lucrative one with tremendous opportunity for big profits but there are also significant risks. Technical analysis is the most useful method of trade forecasting in foreign currency markets. Traders of foreign currency should research past currency prices using charts and trends to properly predict future price performance.

Understanding that history usually repeats itself, technical analysts will need to be able to see familiar patterns in order to make educated decisions regarding future trades.

Using past foreign currency activity to prepare a forecasting model is a tool used in most industries. Although there are tremendous technological breakthroughs, usually the more experience the trader has at chart analysis, the better and more accurate his or her forecasting becomes.

Dow theory is the cornerstone of subsequent chart studies. Charles Dow was basing his research on the stock market alone, at the end of the nineteenth century. Some of the more important points of his theories follow:

- There are three phases to the primary trend: accumulation, run-up & run-down, and distribution. During the accumulation phase the most technically savvy traders begin their market positions. During the run-up/run-down phase most of the market notices the new changes and scrambles to get on board. In the distribution phase, the most proficient traders close out their positions and gather up their profits.

- The market in general has a few trends: minor, primary and secondary. After the primary trend runs its course, the secondary trend will correct the primary trend. The secondary trend will more likely than not retrace part of the way back from the primary trend.

- Confirmation of the trend must be in direct relation to the volume. (Accurate volume figures for currency trading are not readily available, making this theory a tough one to substantiate).

- Currency prices are excellent indicators of all market actions and reactions.

- Trends will continue until reversals of the trends are established as indicated for currency in the charts shown in Figures F.4A and F.4B.

FIGURE F.4A—BULLISH CURRENCY REVERSAL

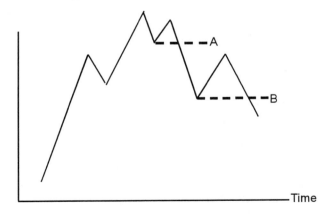

FIGURE F.4B—BEARISH CURRENCY REVERSAL

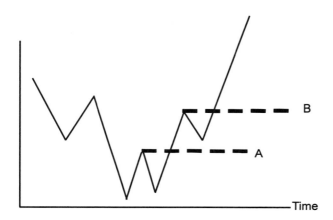

When the technical analyst reviews Figure F.4A, he will note the selling signals at points A and B when the currency dips below previous low points.

He will also notice that the indication to buy in Figure F.4B is indicated at points A and B when the currency rises above previous high points.

See also Futures Trading (Forward Exchange Contract).

CYCLES

In a cycle events, or the tendency for events to repeat themselves, occur at the same time. Cycles are extremely useful in the technical analysis of short and long-term trends in the foreign currency markets.

Cycles look like a wave, the top being the crest and the bottom the trough. A cycle is a measurement from trough to trough, as evidenced in Figure F.5.

FIGURE F.5—THE STRUCTURE OF CYCLES

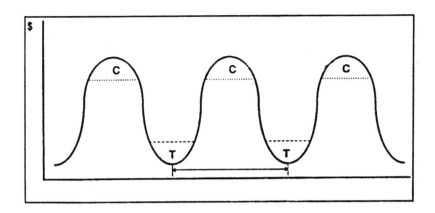

The analysis of a cycle is defined in terms of three indicators (see Figure F.6):

1. *Phase* indicates the point of a wave trough.
2. *Amplitude* illustrates the height of a wave cycle.
3. *Period* indicates the length of the cycle.

FIGURE F.6—THE THREE GAUGING MEASURES OF A CYCLE: PERIOD, AMPLITUDE, AND PHASE

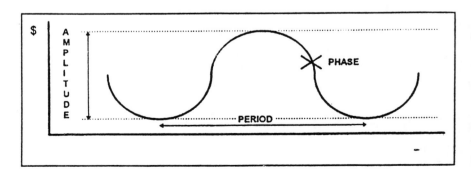

CHARTS

There are many different charts that the technical analyst can use when analyzing foreign currency markets.

The *line chart* (Figure F.7) is the most basic chart, indicating singular prices during a certain time period that are connected by a line.

FIGURE F.7—LINE CHART: AN EXAMPLE

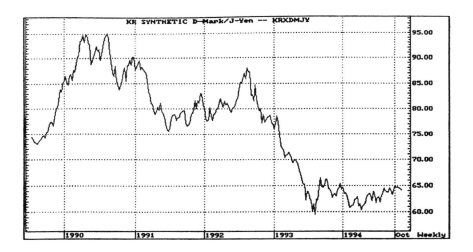

Source: Bridge Information

A *bar chart* (Figure F.8) is a popular chart with several important elements.

- The *price at opening* is indicated by a small horizontal line to the left of the bar.
- The *price at closing* is indicated by a small horizontal line to the right of the bar.
- A vertical bar connects the high and low prices.

FIGURE F.8—BAR CHART: AN EXAMPLE

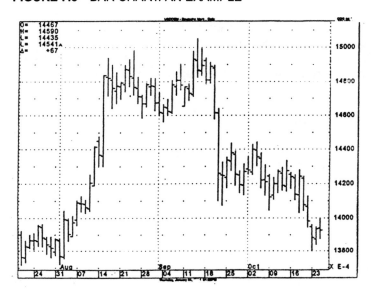

Source: *CQG. Copyright CQG Inc.*

Figure F.9 illustrates the price structure of a bar chart.

FIGURE F.9—PRICE STRUCTURE OF A BAR CHART: AN EXAMPLE

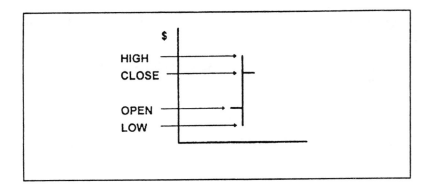

The *candlestick chart* is very closely related to the bar chart; it also shows four different prices—low, high, open, and close--but it is easier to read than bar charts.

Figure F.10 shows the price structure of a candlestick while Figure F.11 presents an actual candlestick chart.

FIGURE F.10—PRICE STRUCTURE OF A CANDLESTICK: AN EXAMPLE

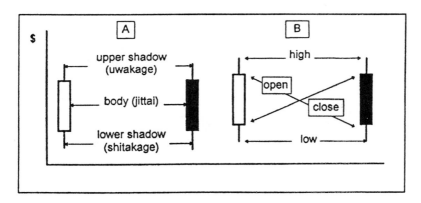

FIGURE F.11—THE U.S. DOLLAR/JAPANESE YEN CANDLESTICK CHART

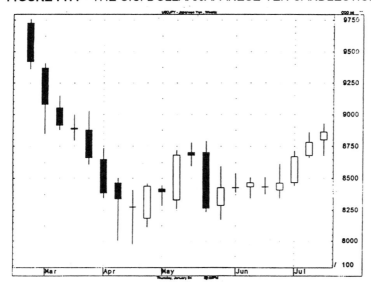

Source: CQG. Copyright CQG INC.

The *point and figure chart* (Figure F.12) does not follow the same approach as the other charts. It does not include the time factor in its analysis, just price activity. Xs indicate upward fluctuations in currency movements. Os indicate downward fluctuations in currency movements.

FIGURE F.12—THE U.S. DOLLAR/SWISS FRANC POINT AND
FIGURE CHART

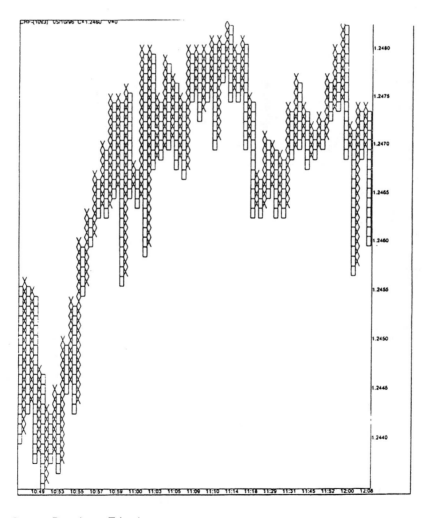

Source: Dow Jones Telerate

CURRENCY FUTURES AND CHARTING

Bar charts are the most popular charts among currency futures traders. Traders will be alert to signals from these charts because currency futures do not actively trade on a 24-hour basis. Chart formations that traders watch in relation to currency futures are gaps, key reversals, and island reversals.

Gaps occur when an opening is outside the previous day's or other period's trading range. Price gaps are significantly common to the currency futures market. Cash is traded around the clock, although foreign currency markets are only actively open for approximately one third of a typical trading day. This leads to day-to-day price gaps in the foreign currency futures market. The trading period in the currency futures trading markets is shorter than in the cash markets so price gaps may occur daily.

A typical bar chart is presented in Figure F.13, a typical price gap in Figure F.14.

FIGURE F.13—BAR CHART: AN EXAMPLE

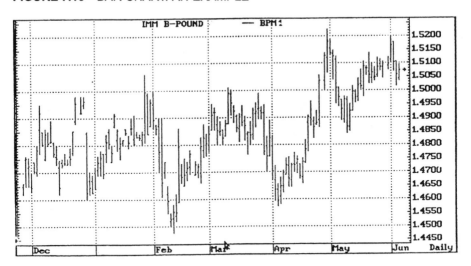

Source: Bridge Information

FIGURE F.14—PRICE GAP: AN EXAMPLE

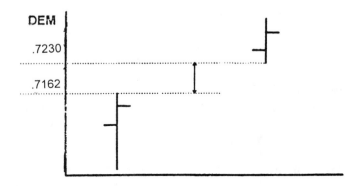

There are a number of different gaps:

Common gaps (Figure F.15) are not technically weighted. They are not good indicators of continuation, reversal, the start of a trend, or the direction of currency in the long term.

FIGURE F.15—COMMON GAPS: AN EXAMPLE

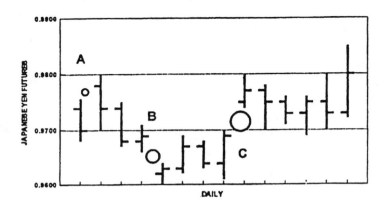

Breakaway gaps (Figure F.16) normally happen at the start of a new trend when there is increased trading and a change in trading mentality.

FIGURE F.16—A TYPICAL BREAKAWAY GAP

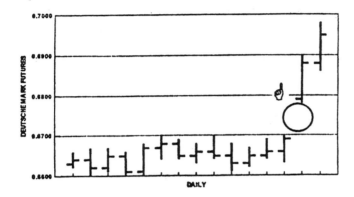

Exhaustion gaps (Figure F.17) normally occur at the bottom or top of a V-formation, as trends change unusually quickly. Following this, there is a sharp trend reversal.

FIGURE F.17—THE PRICE STRUCTURE OF AN EXHAUSTION GAP

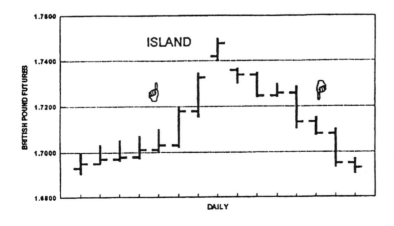

Common trading signals for exhaustion gaps are that:

1. The price objective is not a consideration.
2. The exhaustion gap indicates direction of the market.
3. A new trend draws traders and activity; volume increases.

Island reversals are due to exhaustion gaps. These lines occur at the tips of V-formations where they are separated from the previous and future ranges and are isolated. Usually indicating a strong reversal, they may last for a day or more (Figure F.18).

FIGURE F.18—THE PRICE STRUCTURE OF REVERSAL DAY

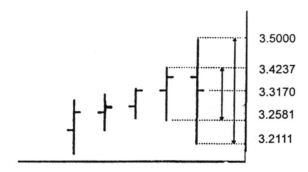

	3.5000
	3.4237
	3.3170
	3.2581
	3.2111

Key reversal day: In the midst of a bullish market a reversal day may happen when a trading range climbs to a new high but closes below the previous day's high. It is a key reversal day (Figure F.19) when the daily price range of the reversal day completely absorbs the previous day's range.

Chart analysis is a reliable tool on which to base trades in the foreign currency market provided the charts are used objectively. Computer software systems and automatic trading systems will make the analytical process much easier and more accurate. Currency traders have diverse interests (e.g., hedging, speculation, long-term vs. short-term positions) requiring different forecasting methods.

FIGURE F.19—THE PRICE STRUCTURE OF KEY REVERSAL DAY

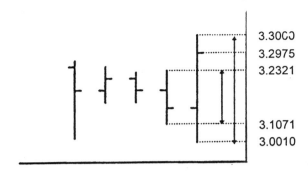

	3.30C0
	3.2975
	3.2321
	3.1071
	3.0010

CURRENCY FORECASTING METHODS

Currency risks may be hedged with currency futures contracts. A currency futures contract is an agreement to buy or sell a specified amount of currency, such as French francs or German marks, at a specified exchange rate by a stipulated future date.

A *forward exchange* transaction is a purchase or sale of foreign currency at an exchange rate set now but with payment and delivery at a specified later date. Most forward exchange contracts have one, three, or six months' maturities. However, contracts for major currencies are available for longer periods, usually one to three years.

In considering currency futures one may wish to consider *floating (flexible) exchange rates*—the movement of a foreign currency exchange rate in response to changes in the supply/demand relationship. The strength or weakness of a foreign currency depends on several factors, including the particular country's inflation rate, the interest rate, economic conditions, the trade balance, and political conditions.

FRONT MONTH

The first available expiration month. For example, if an employee stock option allows the employee to exercise his right starting with January 1, 20X6 and ending with January 1, 20X9, the front month is January 1, 20X6.

FUNDAMENTAL ANALYSIS

Fundamental analysis is the opposite of technical analysis. Fundamental analysis appraises the financial condition of a company, the well-being of the industry the company is part of, and the state of the overall economy to determine the growth potential in firm earnings and stock or bond price. Company analysis should be undertaken after analysis of the economy and industry. For example, it may make less sense to invest in a strong stock in a sick industry than to invest in a weak stock in a very strong industry.

A company's financial ratios are compared to industry averages and to ratios of major competitors. In terms of the economy, consideration is given to budget and trade deficits, inflation rate, production indexes, consumer spending, scarcity of commodities, and interest rates. Plant capacity and management ability are also considered. The company's financial statements are analyzed in terms of trends and ratios. The dividend history of the entity is taken into account. The auditor's report is scrutinized.

Fundamental analysis is in search of the intrinsic value of a stock. Fundamentalists try to uncover the causes of price movement. For example, if the current stock price is below the security's intrinsic value, the decision will be to buy. If a security was fundamentally overvalued, it would not be bought. Of course, the security may be correctly valued.

Note: A deficiency of fundamental analysis is that it involves appraisal of historical information. However, the market is interested in looking ahead.

Some investors combine fundamental and technical analysis. An example is using the fundamental approach to select a particular stock and then using charts for timing decisions on when to buy or sell.

A ratio compares the relative strength of two indicators or measures. For example, the performance of a particular stock or industry can be compared to an overall market measure (e.g., Standard & Poor's 100). How is the individual stock or industry doing compared to the overall market? An analysis of the ratio trend may reveal reversal points. Financial ratios to be examined include liquidity, solvency, profitability, and market value.

Among the ratios and related calculations used in fundamental analysis are:

- *The price/dividend ratio* equals market price per share divided by dividends per share. Although price to dividend is an element of fundamental analysis in appraising a company in absolute terms and relative to other companies, industry averages, and the overall stock market, it can also be looked at from a technical perspective, as in charting the ratio over time. An unusually high ratio may indicate the specific stock or overall market is overvalued; an unusual low ratio points to undervaluation. If the ratio is very high, the decision is to sell because the stock is "expensive"; if low, to buy, because it is cheap. As a general guideline, a very high ratio is above 30, a very low one below 15. This ratio is the opposite of dividend yield.

- *Dividend yield* equals dividends per share divided by market price per share. Stockholders prefer stocks paying a higher dividend relative to price per share.

  Dividend yield can be used to determine if the stock market overall is overvalued or undervalued. It equals total dividends per share for the year for the Standard & Poor's 500 stocks divided by the Standard and Poor's 500 Index. If the dividend yield ratio is low (below 3.4%), stocks are considered overvalued and susceptible to a correction. However, if the rate is high (above 5.2%), stocks are undervalued and represent a buying opportunity.

- *Dividend payout* equals dividends per share divided by earnings per share. Stockholders look unfavorably upon reduced dividends as a sign of possible deterioration of financial health.

- *Volatility ratios*: Volatility is the percentage change (or fluctuation) in price for a stated period (e.g., a week). One measure of volatility is to divide the high by the low for the period.

- *Earnings per share (EPS)* reflect net income per share owned. It is widely watched by investors as a gauge of corporate operating performance and of expected future dividends.

- *Cash per share* is per share cash earnings of the company. Earnings are of higher quality if they are backed by cash because cash may be used to pay debt, buy back stocks or bonds, buy capital assets, pay dividends, and so on.

- *Profit margin* equals net income divided by sales. An entity's profit margin may vary over the years. Profit may be improved by generating additional revenue or by cutting costs. Profit margins are typically higher when there are few companies in a young industry. With increased competition, profit margins will suffer.

  Profit margin is an indicator of operating performance in terms of the firm's pricing, cost structure, and production efficiency. It reveals the entity's ability to generate earnings at a particular sales level. Profit margin may vary greatly within an industry, because it is subject to sales, cost controls, and pricing.

Investors will be reluctant to invest in an entity with poor earnings potential because this adversely affects the market price of stock and future dividends.

- *Return on common equity* is the stockholder's return on the amount invested; it equals net income less preferred dividends divided by the average common stockholders' equity. A company providing a greater return to stockholders in the form of profitability will have a higher stock price.

- *The growth rate* of a business may be expressed in terms of earnings, dividends, sales, market price, and assets. A higher premium is assigned to a company that has a track record of growth.

- *The price-earnings (P/E) ratio (multiple)* equals a company's market price per share divided by its earnings per share.

  Some companies have P/E ratios reflecting high earnings growth expectations. Young, fast-growing companies often have high P/E ratios, with multiples over 20.

  The company's P/E ratio reflects its relationship to its stockholders. It represents the amount investors are willing to pay for each dollar of the company's earnings. A high multiple shows that investors view the firm positively. On the other hand, investors looking for value would prefer a relatively low multiple compared with companies of similar risk and return.

Example: A company's P/E ratio declined from 10 to 8 over the year. The decline in the ratio may have one or more of the following implications:

- Investors have less confidence in the company, perhaps because of the possible deteriorating financial health of the company (e.g., declining growth rate, negative cash flow), increased risk and instability, industry-wide problems, management problems, negative economic situation, or an adverse political environment.

- The company's earnings may have been overstated (poor quality of earnings) and thus are discounted by investors.

  The P/E ratio is typically computed for an individual company but may also be calculated for a group of stocks (e.g., an industry grouping) or for the overall stock market. If the P/E ratio is determined for a group of stocks just add the market prices for all the companies' stocks and divide by the sum of all their earnings per share. If the P/E ratio for the industry group is excessively high, a correction in price may be forthcoming.

A high P/E ratio implies overvaluation, with the possibility that the stock price will decline; a low one implies undervaluation and the possibility of appreciation.

The P/E ratio is used in the fundamental valuation of a company as well as in comparing the company to competing companies, industry averages, and the overall stock market. Most importantly, in terms of technical analysis, the company's P/E ratio should be considered in its market aggregate form. In this regard, according to research done by Ned Davis, if the P/E ratio for the overall stock market is below 10.5, stock prices are undervalued and stocks should be bought; if it is over 18, they are overvalued and should be sold. Hence, the P/E ratio should be looked at in terms of whether market valuations vary from the norm and the likelihood of upward or downward movement to the norm.

The P/E ratio of a company is listed in financial advisory reports and in the financial pages of daily newspapers, notably the *Wall Street Journal*. Industry P/E ratios may be found in financial advisory services such as Standard & Poor's and Moody's.

- *Price-to-book value ratio* is a ratio of market price to book value. Market price of a stock is based on current prices, whereas book value is based on historical prices. Book value per share equals total stockholders' equity divided by total shares outstanding. Market price per share should generally be higher than book value per share as a result of inflation and good corporate performance over the years. A market price below book value could suggest that the company is having financial and operating problems. Some analysts, however, feel a buying opportunity may exist when book value is above market price, because the company stock may be undervalued.

  A company with old assets may have a high ratio, whereas one with new assets may have a low ratio.

Example: A company's share price is $30 and its book value per share is $50. The ratio is 0.6. This company is not doing well. The stock has not kept up with increasing prices. Furthermore, earnings, growth, liquidity, and activity may be deficient. The market may be saying that the assets are overvalued.

The price-book value ratio depends on the industry. For example, many banks have book values that exceed their market price per share.

The price-book value ratio of a company should be compared with that of competing companies within the industry.

- *Price-sales ratio (PSR)* compares the market value of a company's shares to its sales; it reflects a company's underlying financial strength. A company with a low PSR is more attractive while one with a high PSR is less attractive. As a rule of thumb, investors should avoid stocks with a PSR of 1.5 or more, and should sell a stock whose PSR is between 3 to 6.

- *Current ratio* is the ratio of current assets divided by current liabilities. It reflects the company's ability to pay current debt from current assets.

- *Quick (acid-test) ratio* is the ratio of cash plus trading securities plus accounts receivable divided by current liabilities. This is a stringent liquidity indicator.

- The *debt ratio* equals total liabilities divided by total assets. It reveals the amount of money a company owes to its creditors. Excessive debt means greater risk to the investor.

- The *debt/equity ratio* equals total liabilities divided by total stockholders' equity. The ratio shows whether the company has a large amount of debt in its capital structure. A heavily indebted firm has a higher risk of running out of cash in difficult times.

  The fundamental approach similarly looks at the factors impacting on a commodity's price so as to determine the commodity's intrinsic value.

FUTURES TRADING

The future industry started in the U.S. many years ago to guarantee the present commodity price for future delivery. Speculation and hedging practices began as the futures market evolved.

A *futures contract* is an agreement to buy or sell a given amount of a commodity or financial instrument at a specified price in a specified future month. The seller of a futures contract agrees to deliver the item to the buyer of the contract, who agrees to purchase the item. The contract specifies the amount, valuation, method, quality, month, means of delivery, and the commodity exchange to be traded in. The month of delivery is the expiration date when the commodity or financial instrument must be delivered. In commodities futures, price is set by open outcry on the floor of the commodity exchange. The contract specifies the item, price, expiration date and a standardized unit to be traded (e.g., 10,000 pounds). In looking at the change in commodity prices for a particular period, consideration should be given to the total number of contracts traded (volume). Commodity contracts may run up to one year. Investors must continually evaluate the effect of market activity on the value of the contract. While the futures contract mandates that the buyer and seller exchange the commodity on the delivery date, the contract may be sold to another party before the settlement date. This may occur when the trader wants to realize a profit or limit a loss. Investors engage in commodity trading in the hope of high return rates and inflation hedges; they rarely take delivery of the commodity itself.

Interest rate futures are contracts where the holder agrees to take delivery of a given amount of the related debt security at a later date (usually not more than three years). Futures may be in Treasury bills and notes, CDs, commercial paper, or GNMA certificates. Interest rate futures are stated as a percentage of the par value of the underlying debt security.

The value of these futures contracts is directly tied to interest rates: as interest rates decrease, the value of the contract increases. As the price or quote goes up, the purchaser of the contract has a gain, while the seller loses. A change of one basis point in interest rates causes a price change.

Those who trade in interest rate futures do not usually take possession of the financial instrument. In essence, the contract is used to hedge or speculate on future interest rates and security prices. Although there is significant risk, speculators find financial futures attractive because of their potentially larger return on a small investment (deposit requirements are low).

Financial instruments are not just cash but also forward contracts, accounts receivable and payables, equity instruments, options and guarantees.

Forward exchange contracts are agreements to exchange at a given future date currencies of different countries at a specified rate (the forward rate). A forward contract is a foreign currency transaction that impacts the profit & loss statement of the company in question.

The technical analyst should realize that his analysis tools are useful on any of the actively traded commodity futures. He should also consider other factors, such as psychological and political ones, as they relate to the price of a commodity futures contract. Some of the methods and tools the technical analyst has at his disposal are figure and point charts, volume, trend lines, moving averages, and bar charts.

The price analysis of commodities is complicated. Each commodity has a different way of measuring price structure, e.g.:

- Cotton is measured in cents per bushel.
- Platinum is measured in dollars per ounce.
- Gold is measured in dollars per ounce.

Therefore, there is no set yardstick to use when comparing various price structures. The analyst must understand the contract rules of each market, how the contract price is arrived at, and what the maximum and minimum price measures are.

Commodity futures contracts have a limited life span and defined expiration dates. The normal futures contract trades for approximately 18 months before expiration. Long-range price forecasting is difficult. The technical analyst must realize the dynamics of commodity futures and take appropriate measures to keep charts and calculations current. New historical data is needed as the outdated contracts expire, thus making the amount and depth of technical data increasingly difficult to keep current.

Substantial amounts of money can be made or lost with great speed in trading futures. They are usually traded on margin that is less than 10% of the contract value. Substantial leverage is provided by these low margin requirements.

Example:

Value of futures contract	$500,000
Margin percentage	10%
Trader's margin requirement	50,000
Increase in futures contract (10%)	$550,000
(Trader doubles his money)	
Decrease in futures contract (10%)	$450,000
(Trader loses his investment)	

A 10% move can either double the trader's initial investment or eliminate it completely. Accurate timing and quick response to exit and entry opportunities due to the increased leverage factor are paramount for success in this market.

The Commodity Research Bureau (CRB) Price Index is a barometer of commodity prices that measures the trend performance of 27 commodity futures markets. The Attitude Index determines how many of the 23 physical commodities followed by Market Vane are bullish; if more than 50% are bullish, it may indicate an upcoming trend.

TRENDS AND TREND FOLLOWING SYSTEMS

The futures market is dynamic. Futures traders have a more short-term outlook than traders in other markets. Futures traders will use various technical analysis techniques aimed at the short term, one of their favorites being the moving average.

Moving average is an average of a specific group of data for a specific period and only for that period. If it were to be measured for an 8-day period it would be constantly moving forward, taking into account only the immediate past 8 days.

Moving average measures are part of the larger trend following system. Usually, a trend following system will not buy near the low points or sell near the high ones, requiring an opposite shift in price to make a trade.

The moving average is tremendously versatile because it is easy to measure and test. Moving average calculations are easy to measure via computer analysis, unlike chart systems, which are difficult to computerize. The moving average analysis will indicate buy and sell signals that the technical analyst can rely on.

Timing is of paramount importance in futures trading. Technical analysts need to know the direction the market is heading and the optimum point at which to jump in. Timing is so critical that being off by even a few minutes can create a loser.

In general, a *trend* is the direction in which the market is moving, either up or down. Normally, markets do not move in straight lines. They move in a zigzag manner characterized by waves that form a series of peaks and valleys. The direction of the peaks and valleys shapes the market trend. These peaks and valleys can move in any direction, up, down, or sideways.

A *downtrend* is a series of successively lower peaks and valleys while an uptrend is a series of successively higher peaks and valleys. A sideways market trend is indicated by a series of horizontal peaks and valleys. This type of market is often referred to as trendless. (See Figures F.20, F.21, and F.22.)

FIGURE F.20—AN UPTREND WITH ASCENDING PEAKS AND TROUGHS

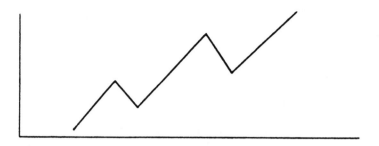

FIGURE F.21—A DOWNTREND WITH DESCENDING PEAKS AND TROUGHS

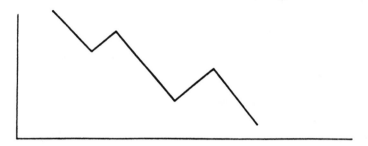

FIGURE F.22—A SIDEWAYS TREND WITH HORIZONTAL PEAKS AND
TROUGHS

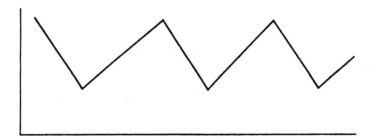

Uptrends and downtrends are sometimes described in terms of *trend lines*. Trend lines can extend for long periods of time. A downtrend line connects a series of lower highs; in contrast, an uptrend line connects a series of higher lows. Trend channels are parallel lines that enclose a trend. (See Figures F.23, F.24, and F.25.)

FIGURE F.23—UPTREND LINE: JULY 1993 SILVER

FIGURE F.24—DOWNTREND LINE: COCOA CONTINUOUS FUTURES

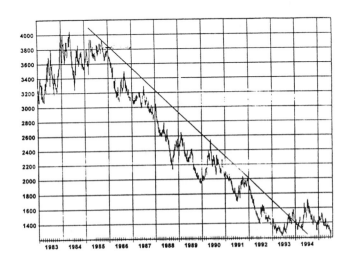

FIGURE F.25—DOWNTREND CHANNEL: SEPTEMBER 1992 COCOA

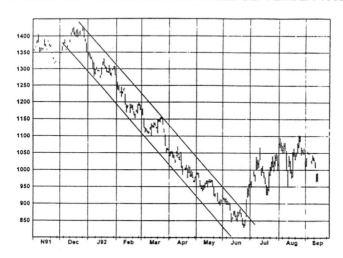

The technical analyst should be aware of the following indicators:

- For the short-term trader, the bottom end of a downtrend channel and the upper end of an uptrend channel indicate possible profit areas.

- Form a position in the direction of a major trend when there are declines approaching an uptrend line and there are inclines approaching a downtrend line.

- Sell signals are indicated by the penetration of an uptrend line, a buy signal by the penetration of a downtrend line. Penetration is indicated by a minimum amount of closes past the trend line or a minimum percentage price move.

There are many other trend-following systems, including breakout systems, fading minimum move, oscillators, cycles, and even contrary opinion. There are also some common problems with regard to trend-following systems, but thankfully, there are some logical approaches:

Problem	Approach
Large number of similar systems	Don't trade with everyone else
	Implement original system

Problem	**Approach**
System works great on paper but fails in a practical arena	Test system properly
Sluggish systems can give up large profits	Use trade exit rules

TRADING RANGES

The trading range is a horizontal channel where prices vary for an extended time. Trading ranges are extremely difficult to predict, and it is difficult to make a profit. Most traders would be wise to limit their involvement in these markets. Breakouts from a trading range indicate a price fluctuation in the direction of the breakout. (See Figures F.26 and F.27.)

FIGURE F.26—UPSIDE BREAKOUT FROM TRADING RANGE: DECEMBER 1993 T-BOND

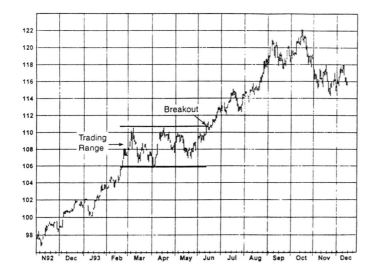

FIGURE F.27—DOWNSIDE BREAKOUT FROM TRADING RANGE: CATTLE
CONTINUOUS FUTURES

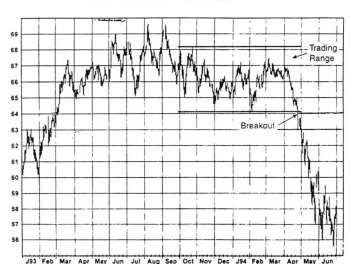

Support and Resistance

When the technical analyst notices a one- to two-month price fluctuation in a sideways trading range, prices will usually meet support at the lower end and resistance on the upper end.

Usually, after prices jump out from a current trading range, support and resistance definitions no longer apply. If prices experience a continued breakout above a trading range, the upper limits of that range will be a price support zone.

Resistance will usually be met with where there are previous major highs and support in the area of major lows. An important thing to consider is that a previous high does not guarantee that future rallies will falter at or below that point. Near that point resistance should be expected.

The converse is also true: A past low does not insure that future declines will stay at or above that point or that support should be expected in the general area of that point (see Figure F.28).

FIGURE F.28—RESISTANCE AT PREVIOUS HIGH AND SUPPORT AT
PREVIOUS LOW: SOYBEAN OIL NEAREST FUTURES

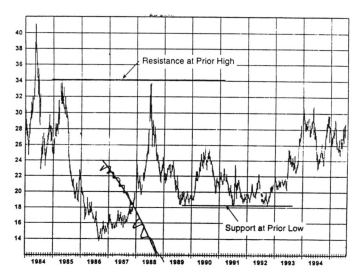

Pyramiding and Mid-trend Entry

When there is a major trend going on, how do you enter the market? You
must control your risk and have great timing. The trader could:

- Take advantage of the natural tendency of the market to follow
  previous price swings, watching the market for the last relative
  high or low and starting a position when the market duplicates a
  given percentage of that price swing. This method gives the trader
  excellent entry points; sometimes the duplication will not occur
  until the trend has gone on for much longer.

- Wait for a minor reaction to occur and for the resumption of a
  major trend, and then start a position.

Pyramiding is adding units to an existing market position. The issue
of pyramiding as it compares to mid-trend entry is that the trader initiates
a position after the market has moved in a given direction. When imple-
menting pyramiding, do not increase a position until the last unit posi-
tioned shows a profit. Also, do not increase the position if your stop point
would indicate a net loss.

Pyramid units should be no more than the original position size.

Stop/Loss Points

Success in trading in futures depends on controlling losses. The exact stop-loss point should be calculated before beginning a trade. One basic principle: The position should be sold out before or right at the instance at which price movement causes a shift in the technical picture (see Figure F.29).

FIGURE F.29—STOP PLACEMENT FOLLOWING TRADING RANGE BREAKOUT: DECEMBER 1994 T-BOND

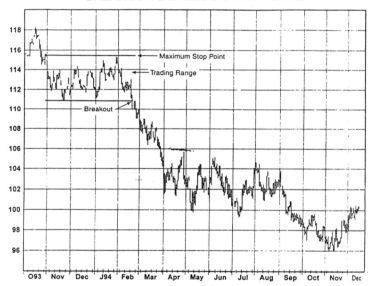

There are various technical reference points at which protective stop orders can be placed:

Trend lines: A purchase stop can be implemented above a downtrend line and a sell stop below an uptrend line. The major advantage of this method is that the infiltration of a trend line is normally the initial technical signal indicating a trend reversal.

Flags and pennants: Usually after a breakout in one direction of a pennant or a flag arrangement, a return to the opposite end or further signifies a price reversal—and the place for a stop order.

Relative lows and highs: The most current relative high or low can be used as a technical indicator to place a stop point.

Expanded days. After a breakout in one direction, the return to the opposite side should indicate a price reversal and a good time to place a stop order.

ANALYSIS OF CYCLES IN THE FUTURES MARKETS

The study of cycles is extensive. Technical analysts are concerned with periodic cycles occurring at regular intervals with recurring patterns in data. (See Figure F.30.)

FIGURE F.30—IDEAL CYCLE MODEL

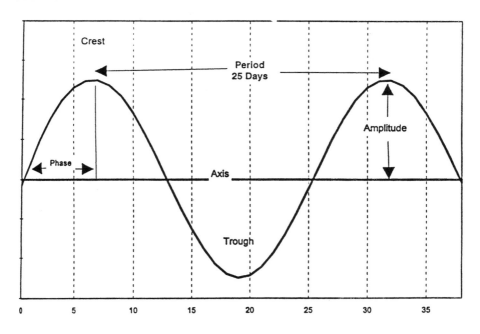

Traders often make the mistake of relying implicitly on cyclical data and expecting that market upswings and downswings will conform exactly to the cycles found. There are related problems, including the fact that *swings in market trading are not symmetrical.* The curves used to explain cycles are symmetrical, taking for granted that downswings and upswings in the market last for the same length of time. Market swings are usually translated. If downswings last longer than upswings, the cycle has a left-hand translation (bearish). A right-handed translation is bullish.

Cycle analysis is based in large part on past data and occurrences. When cycle crests and troughs are early or late and do not match up to past performances taken a referential basis, there is a problem. An example of cycle translations is shown in Figure F.31.

FIGURE F.31—CYCLE TRANSLATIONS

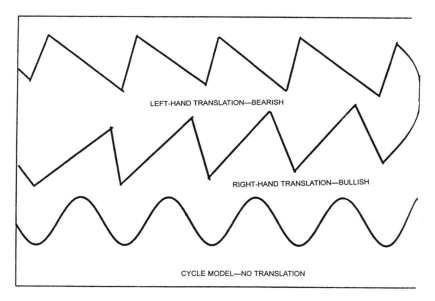

The astute technical analyst should note the positioning of the previous peaks in a cycle for insights into the direction of translation in future cycles.

Inexperienced traders often try to pick market tops and bottoms based on cyclical analysis alone. The well-informed technical analyst will use all methods at his disposal to make accurate trading decisions.

GETTING THE MOST OUT OF TRADING SYSTEMS

A trading system is simply a set of parameters that indicate trade signals. A trading system needs to be tested over an extended period within a wide series of markets before it is used in earnest. Most trading systems have a number of parameters that will indicate a trade; most general systems are limited to one or two parameters. It is usually a good idea to use the simplest system with the least amount of parameters.

Usually, the longer the test period, the more reliable the results. If the period is not long enough, it will not properly measure performance of the trading system in a comfortable range of market situations. If it is too long, previous periods represented may not be indicative of current market conditions.

Sometimes traders discover that actual results differ from the paper trading results obtained by their system, a common situation called slippage. Assuming there are no errors in the program, slippage results from failure to use realistic assumptions when testing the system.

PLANNED TRADING

Success at trading in the futures market requires a disciplined and systematic trading approach. In order to achieve success, you must adhere to some general guidelines:

1. *Adopt a trading philosophy.* A sound strategy would be based on trading systems with stop-loss points clearly defined and basic chart analysis.

2. *Define the markets you will trade in.* Usually the funds a trader has available will dictate the type and number of markets he will trade in. A good rule of thumb for a trader with limited funds should be to avoid volatile markets. This will generally restrict the number of markets that can be traded in.

3. *Develop a hazard control plan.* A loss control plan is of paramount importance: You must have a clear definition of your exit point before you begin trading.

4. *Keep accurate trading records.* A beginner should trade on paper to get used to the trading philosophy he will employ before trading for real. This suggestion is important to get a feel for the discipline involved in common sense trading before using actual funds. Once you start trading with real money, keep good records so you can evaluate the effectiveness of your philosophy.

5. *Get in the habit of planning your strategy.* No trader should do any trading without an intelligently thought-out plan. Set aside time for review of the previous periods activity with an eye fixed on the future.

6. *Be honest about your trading ability.* Analyze your strengths and weaknesses and your overall approach to trading. Is your trading philosophy a sound one? The answer can be found by examining past successes or failures.

The most important thing a trader can do is to learn quickly from mistakes. Repeating the same trading errors before a new strategy is developed can lead to tremendous losses.

G

GAPS

Areas on a bar chart where no trading has occurred. Sometimes, prices open higher than the top price of the previous day, leaving an unoccupied space—a gap—on the graph. Inequities in supply and demand are the cause of these gaps. A gap can also be explained by a price range where no trading has occurred or an upward swing when the top price of day one is below the lowest price of day two. A gap in a downswing can result when day 1's lowest price is above day 2's highest price.

Some gaps can be closed by a price change or movement. Some technical analysts believe or assume that prices must close these gaps, but there is really no clear indication as to when that will occur. Closing (covering) the gap occurs when a stock or commodity returns to a previous gap and retraces its range. Figures G.1 and G.2 depict gaps in an uptrend and a downtrend:

Gaps, by themselves, should not be relied on as indicators of financial movement. Though some gaps are important, others have no significance at all. Gaps are not apparent on all charts. They do not reveal themselves on a point and figure chart; they do show up on a bar chart. Inconsequential gaps happen often with thinly traded stocks because they result from infrequent trading. In time, most gaps will be closed, though there is really no reason why they should be.

The following are examples of a gap being closed after time has elapsed. As can be seen by these examples, making investment decisions that rely on the theory that gaps must be closed is not a good investment strategy.

FIGURE G.1—GAPS IN AN UPTREND

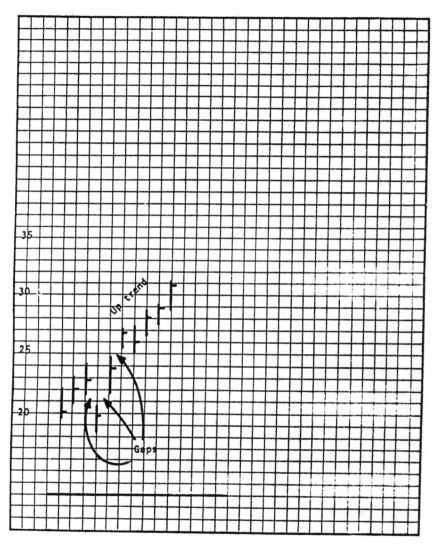

FIGURE G.2—GAPS IN A DOWNTREND

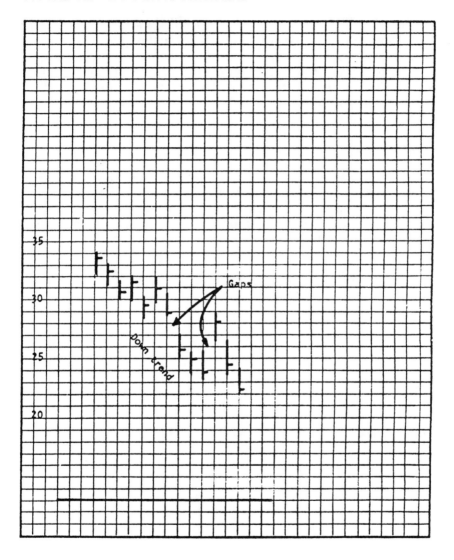

Example 1:

Over a period of nine months Roberta Corporation's stock climbed from $22 per share to $45 per share. One day it closed at the high of its range—$45. The next day the stock opened at $45½ and moved upward, causing a gap. The price of the stock rose to $54 before coming back down to $45½. The gap is now closed but only after nine months have elapsed. A price gap may bypass a price range because of some unusual news, positive or negative, about the company.

Example 2:

On day 1 a stock price got no higher than $12. The next day it opens up in the price range of $15-$16 on news of a takeover bid.

    Gaps usually occur more often on charts that cover short time spans (daily) than on charts that track longer time frames such as those that encompass monthly data. Figure G.3 depicts a typical gap.

FIGURE G.3—GAPS

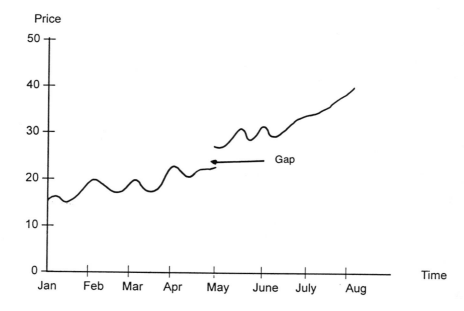

TYPES OF GAPS

The types of gap include common, breakaway, runaway, reversal, exhaustion, and ex-dividend.

A *common gap* usually occurs in thinly traded markets or where prices are congested (e.g., midway in a horizontal trading range). This is an insignificant type of gap that shows disinterest in trading. The gap is a hole in a chart occurring within an area pattern (see Figure G.4).

FIGURE G.4—COMMON GAPS

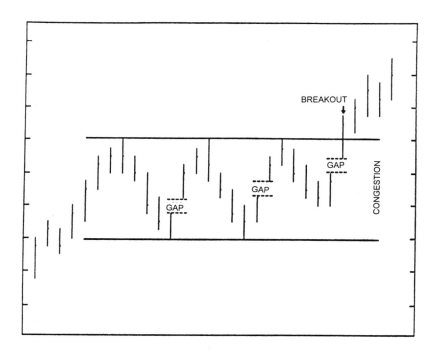

A *breakaway gap* usually happens toward the end of a price pattern, before prices break away. Breakaway gaps normally operate in conjunction with upside breakouts and are not typically recorded as often as they occur because they are usually intraday. These gaps are good indicators that a price breakout is ahead (see Figure G.5). Prices will move faster as soon as the breakout occurs due to heavy buying action. Buyers will usually have to make increased efforts to find sellers.

Tip: If you own or short a stock that is in a pattern formation and the stock experiences a breakaway gap in an adverse direction, close the position.

FIGURE G.5—BREAKAWAY GAP

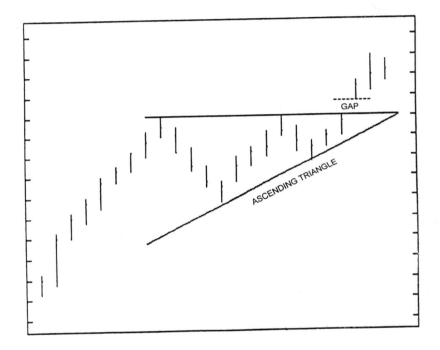

The *runaway gap* is an excellent indicator of a powerful underlying trend. Runaway gaps are noticeable in the center of fast up or down price fluctuations; they happen often where price quotes are moving rapidly relative to transaction volume. Unlike breakaway or common gaps, runaway gaps are important from a technical standpoint because runaway gaps are normally not filled until the next significant price move. Usually, a substantial amount of time will have passed. A runaway (measuring) gap often takes place about midpoint in a key market trend. If stock price is increasing significantly and there is a runaway gap, hold the stock until it rises to the gap.

A *reversal gap* is a chart formation where the price bottom of the day's trading is higher than the price range of the previous day. The closing price is greater than the opening price and higher than the midrange price.

The *exhaustion gap* (Figure G.6) is the final showing of an increasing or decreasing price move, indicating that a trend is ending. Very similar to a runaway gap, it is normally linked to a quick up or down price move at the culmination of a major trend. Prices are usually moving up until they encounter a large supply in an upward movement, or considerable demand on the downward slide. When this occurs the movement is quickly ended by a high volume trading day.

An exhaustion gap typically appears between the last day and the next-to-last day of a price move. If there is an exhaustion gap after a solid increase in stock price, protect your position with a close progressive stop.

Occasionally after an upward exhaustion gap, prices trade in a narrow range for a week or several days before gapping to the downside. This very short period of time makes the price action resemble an island.

FIGURE G.6—EXHAUSTION GAP

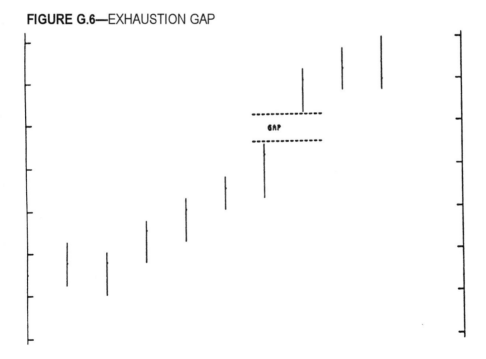

The *ex-dividend gap* reflects the ex-dividend date; the day the dividend reduces the market price of a stock.

Any of these price gaps may occur when a stock price either jumps or plummets from its last trading range but does not overlap that trading range. Price gaps are studied by technical analysts because the gaps may point to overbought or oversold conditions.

The various types of gaps can appear in any order. Their significance can range from very low (common) to great (exhaustion).

If you own a stock that is in a clear pattern or in an area within narrow limits, a sudden positive breakout of the stock with significant volume and a gap is a bullish sign. Stock bought at this time should probably be held until exhaustion appears.

In conclusion, an up gap is seen when the lowest price for a trading day is higher than the highest high of the previous day. An up gap indicates strength in the market. A down gap indicating market weakness is seen when the highest price for the day is lower than the lowest low of the previous day.

HAMMERING THE MARKET

Intense selling of securities by investors who believe stock prices are excessive. Speculators expecting a market decline sell short, thus hammering stock prices.

HEAD AND SHOULDERS

A pattern (Figure H.1) in which a stock price goes to a peak, declines, and then increases to a second peak higher than the first peak, declines again, and then goes to a third peak that is lower than the second peak. The pattern, which resembles a head with a left and right shoulder, indicates the reversal of a trend.

The *neckline* is a line drawn through the two bottom points of the two valleys between the head and shoulders of a head and shoulders pattern.

A reverse head and shoulders pattern has the head at the bottom of the chart, meaning that prices should be increasing.

HERRICK PAYOFF INDEX (HPI)

Developed by John Herrick, the Herrick Payoff Index (HPI) is designed to show the amount of money flowing into or out of a futures contract. The index uses open interest during its calculations; therefore, the security being analyzed must contain open interest. An HPI above zero shows that money is flowing into the futures contract (bullish). An HPI below zero shows that money is flowing out of the contract (bearish). The HPI is interpreted by looking for divergences between the index and prices.

FIGURE H.1—HEAD AND SHOULDERS FORMATION

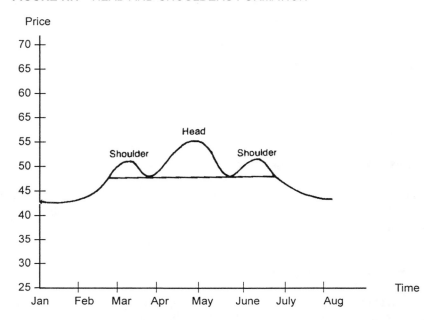

Example:

Figure H.2 shows the British pound and the HPI. The trendlines identify a bearish divergence where prices were making new highs while the HPI was not. As is typical with divergences, prices corrected to confirm the indicator.

To compute the HPI:

$$\frac{Ky + ((K' - Ky) * S)}{100{,}000}$$

where:

Ky = The previous period's HPI

$$K' = (C* V*(M - My)) \left[1 \pm \frac{2 * I}{G} \right]$$

S = The multiplying factor

C   = The value of a one cent move

V   = Today's volume

$$M = \frac{High + Low}{2}$$

My   = M (mean price) yesterday

+   = Replace the "+" symbol with "+" if M is greater
          than My or with "-" if less than My

I   = The absolute value of the difference
          between today's open interest and
          yesterday's open interest

G   = The greater of today's or
          yesterday's open interest

FIGURE H.2—HERRICK PAYOFF INDEX: BRITISH POUND

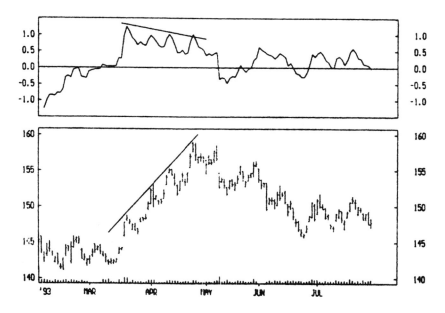

The HPI requires two inputs, a smoothing factor known as the multiplying factor and the value of a one-cent move. The multiplying factor produces results similar to the smoothing obtained by a moving average. For example, a multiplying factor of 10 produces results similar to a 10-period moving average. The recommended value of a one-cent move is 100 for all commodities except silver, which should be 50.

HISTORICAL TRADING RANGE

The price range within which a stock, bond, or commodity has traded since going public.

HOOK DAY

A trading day experiencing an open that is higher than the previous day's high or lower than its low while the close is lower in the first case or higher in the second than the previous day's close.

I

INDEX OF BEARISH SENTIMENT

A contrary opinion rule developed by A.W. Cohen of Chartcraft that is based on a reversal of the recommendations of investment advisory services as contained in their market letters. The index equals:

$$\frac{\text{Bearish investment advisory services}}{\text{Total number of investment advisory services}}$$

A ratio of 42% or more indicates that the market will go up because it is a contrary indicator; a ratio of 17% or less suggests the market will go down.

Example:

Of 200 investment advisory services, 90 are bearish on the stock market. The index equals 0.45 (90/200). Since 45% are pessimistic, or more than the 42% benchmark, the investor should be buying.

The investor should use this index in predicting the future direction of the securities market based on contrary opinion. If sentiment is bearish, a bull market is expected, and it is time to buy. If sentiment is bullish, a bear market is likely, and it is time to sell securities.

The Index of Bearish Sentiment, which is published by Investors Intelligence, can be found in *Barron's*.

INSIDER TRADING INDICATOR

The Securities and Exchange Commission defines an insider as any officer or director of a corporation or any entity that owns more than 10 percent of a company's stock. The insider trading indicator plots a graph of historical trades made by insiders of a corporation. The indicator is based on the number of separate insiders trading over a given period. For example, one insider selling 100 shares is one trade and one insider selling 1,000 or 10,000 shares is also one trade. It does not weight the number of shares traded. Thus, net trades are plotted: One net trade for one month equals insider buys minus insider sales. The inclusion of buys, exercise of options, and sells can be set in the legend.

Heavy net insider buying is considered to be positive for a stock, while heavy net selling is considered negative. Historical testing shows that insider buying is more accurate than insider selling for foretelling future price movements; stocks with heavy insider buying outperform the market over long periods of time.

Figure I.1 displays inside trading activity for Microsoft.

INSIDERS

An insider is a director, officer, or beneficial owner of a company's stock who may have access to key information before it is announced to the public. An insider is extremely knowledgeable about corporate sensitive information not available to the average investor, including, for example, knowledge of a forthcoming takeover of the company, or of surprises in an earnings report yet to be released. Concentrated buying could take place before good news is revealed to the public. Concentrated insider selling might reflect knowledge of a change in the fundamental picture not yet recognized by the public.

That is why insiders are not allowed to take advantage of any information before it is released to the public. Further, insiders are prohibited from profiting on their transactions until at least six months after the transaction. Any stockholder owning more than 5% of company voting common stock must inform the Securities and Exchange Commission when he buys or sells shares. Because insiders are privy to a lot of financial and other information about the company, the SEC rules are designed to prevent abuse. Insider violations can be punished by fine and imprisonment, among other remedies.

If a company is publicly held, the insider must report any buys or sells of the company's shares to the Securities and Exchange Commission within the first 10 days of the month subsequent to the month of the transaction.

FIGURE I.1—INSIDER TRADING: MICROSOFT MARCH 1998-FEBRUARY 1999

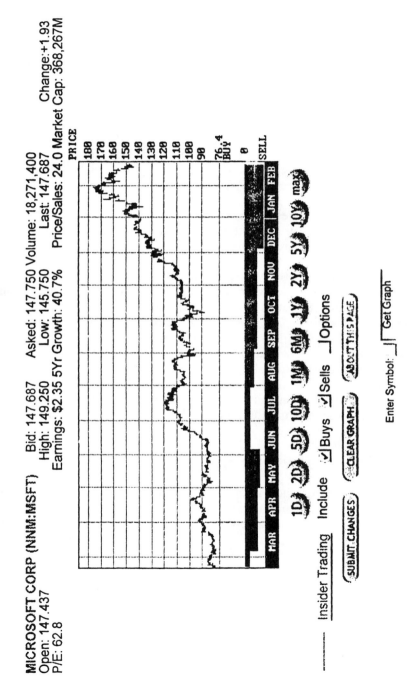

MICROSOFT CORP (NNM:MSFT) Bid: 147.687 Asked: 147.750 Volume: 18,271,400 Change: +1.93
Open: 147.437 High: 149.250 Low: 145.750 Last: 147.687
P/E: 62.8 Earnings: $2.35 5Yr Growth: 40.7% Price/Sales: 24.0 Market Cap: 368,267M

Current Stock Quotes are delayed a minimum of 20 minutes – 02/19/99 - 5:02 p.m. Eastern

Insiders may legally buy when they believe the stock is undervalued and sell when they think it is overvalued. In other words, insiders as a group usually accumulate more shares at market bottoms and do more selling around market tops. Insiders are usually correct in their timing decisions. For this reason investors should monitor transactions made by insiders.

An insider sell/buy ratio is prepared by Vickers Stock Research Corporation.

A high sell/buy ratio points to stock overvaluation, a low ratio to undervaluation. The bottom line is to follow what insiders are doing in their buying or selling patterns. If the ratio is 1.5 or less, buy. If the ratio is 2 or more and the stock market is down 5% from the buy level or 5% from a later high, sell.

There is a timeliness problem because insider information may be several months old by the time it is released. Therefore, the investor's timing is crucial or there will be little benefit to knowing insider transactions.

Insider trading activity is published in the SEC monthly *Official Summary of Security Transactions and Holdings* and in newsletters and financial publications such as *Invest Net, Consensus of Insiders (COI), The Insider Indicator, The Insiders, and Vickers Weekly Insider Report.* For example, the COI lists the 20 most attractive stocks. These private services make recommendations based on insider buying and selling. *The Insider Chronicle* (editorial section of the *Commercial and Financial Chronicle*) publishes a listing of insiders' transactions for about 400 companies. Other sources include *Barron's, Insider Trading Monitor* (a database service that reports SEC filings), and the *Wall Street Journal*.

INTERDELIVERY SPREAD

A futures or options trading approach that involves buying one month of a contract and selling another month in the same contract hoping to profit from the widening or narrowing of the price difference between the contracts.

Example: An investor simultaneously buys an August cotton contract and sells a November cotton contract.

INTEREST RATES

DISCOUNT, FED FUNDS, AND PRIME

These are three key interest rates closely tied to the banking system. The *discount rate* is the rate the Federal Reserve Board charges on loans to banks in the Fed system. The federal or fed funds rate is the rate bankers charge one another for very short-term loans, although the Fed can manage this rate as well. The *prime rate* is the benchmark rate that bankers charge their best corporate customers. The three rates at times work in tandem.

The discount rate is the major tool that the nation's central bank, the Federal Reserve, has to manage interest rates. The Fed changes the discount rate when it wants to use monetary policy to alter economic patterns.

The discount rate can instantly impact the fed funds rate, which is more market-driven and changes throughout each business day. Although it can at times be managed by the Federal Reserve, it also can move as bankers and traders anticipate Fed activity.

The prime rate is a heavily tracked rate although it is no longer as widely used as a loan benchmark as it has been in the past. The prime is set by bankers to vary loan rates to smaller businesses and on consumers' home equity loans and credit cards.

Changes in the discount rate and prime rate are often front-page news, particularly in times of economic troubles. Many major newspapers such as the *Baltimore Sun*, the *Miami Herald*, and the *Los Angeles Times* carry a daily rate tally that includes the discount, fed funds, and prime rates.

These three interest rates reflect the banking system's view of the strength of the U.S. economy.   The rates tend to rise when the Federal Reserve is in a tight money posture trying to keep an expanding economy from overheating and creating too much inflation. Their rise is often viewed as a negative signal for both stock and bond investments, because a cooled economy and higher rates can hurt many securities prices.

Conversely, the rates tend to fall when the Federal Reserve is in an easy money mode—when it wants available credit and low interest rates to stimulate a moribund U.S. economy. For the risk-taking investor, this can be a signal to boost stock and bond holdings. An expected economic recovery may boost equity issues, while falling rates should be a boon to bond prices.

At times, when bankers and the Federal Reserve disagree on the direction of the economy, these rates may send contradictory signals.

30-YEAR TREASURY BONDS

The most widely watched interest rate in the world, the security known as the "T-bond" is seen as the daily barometer of how the bond market is performing. The 30-year Treasury bond is a fixed-rate direct obligation of the U.S. government.

Traders monitor the price of the U.S. Treasury's most recently issued 30-year bond, often called the bellwether. The price is set by dealers who own the exclusive right to make U.S. markets in the bonds. (The bond trades around the clock in foreign markets.) Bond yields are derived from the current trading price and its fixed coupon rate.

The T-bond price and yield can be found in the credit market wrap-up in newspapers such as the *Boston Globe*, *Investor's Business Daily*, the *New York Times*, and the *Wall Street Journal*, and in computer databases such as Prodigy.

Traders who hold T-bonds are exposed to a great deal of risk. Their degree of willingness to take on that risk—and the changes that it brings to 30-year bond prices and yields—is often viewed as a proxy for the long-term outlook for the U.S. economy.

Because it is long term, the T-bond is extremely sensitive to inflation that could ravage the buying power of its fixed-rate pay-outs. Thus, the T-bond market also is watched as an indicator of where inflation may be headed.

Also, T-bond rates directly impact fixed-rate mortgages. Consequently, they also are seen as a barometer for the housing industry, a key leading indicator for the economy.

THREE-MONTH TREASURY BILLS

The Treasury bill rate is a widely watched rate for secure cash investments. In turbulent times the rate can be volatile.

T-bills, both 3-month and 6-month issues, are auctioned every Monday by the U.S. Treasury through the Federal Reserve. Rather than paying interest, the securities are sold at a varying discount depending on the prevailing interest rate. Investors get their interest by redeeming the T-bill at full face value when it matures.

The government reports the average effective interest rate it pays each week.

Results of the Monday auctions can be found in the business sections of most major daily newspapers, such as the *New York Times* and the *Wall Street Journal*. Trading in the secondary markets for T-bills also is reported there as well as in most major daily newspapers and on business TV channels such as CNBC. Barron's reports it as shown in Figure I.2.

FIGURE I.2—MONEY RATES REPORT

	Latest Week	Prev. Week	Yr. Ago Week
Discount Rate (NY)	4 1/2	4 1/2	5
Prime Rate (base)	7 3/4	7 3/4	8 1/2
Fed Funds Rate			
Avg effective offer	4 7/8	4 7/8	5 7/16
Avg weekly auction-c	4.75	4.66	5.52
T-Bills Rate			
13 weeks, Coupon Yield	4.531	4.51	5.234
13 weeks, Auction Rate	4.42	4.40	5.095
26 weeks, Coupon Yield	4.585	4.551	5.282
26 weeks, Auction Rate	4.42	4.39	5.075
52 weeks, Coupon Yield	4.584	4.584	4.965
52 weeks, Auction Rate	4.37	4.37	5.232
Avg weekly auction-c	4.40	4.35	5.05
Broker Call Rate			
CD's Rate			
3 months	4.74	4.72	5.23
6 months	4.97	4.96	5.53
Commercial Paper Rate			
Dealer-placed			
1 month	4.83	4.80	5.51
2 months	4.83	4.79	5.50
3 months	4.83	4.79	5.48
Directly-placed (GE Capital)			
30 to 36 days	4.81	4.82	5.47
200 to 270 days	4.77	4.75	5.32
Bankers Acceptances			
1 month	4.76	4.78	5.41
2 months	4.75	4.77	5.41
3 months	4.75	4.77	5.37
6 months	4.75	4.75	5.32
Libor Eurodollar Rate			
3 months	5.00	4.9825	5 5/8
6 months	5.03531	5.03031	5 5/8
12 months	5.18063	5.15359	5 11/16

Source: *Barron's*, February 15, 1999

The T-bill rate shows what no-risk investments can be expected to earn. Historically, T-bills have returned little more than the inflation rate. Many conservative investors buy T-bills directly from the government. T-bill rates approximate rates on money market mutual funds or statement savings accounts, also popular savings tools for the small investor.

When these low-risk rates are high or rising, it can be a negative signal for stocks and bonds because individual investors are shying away from riskier investments.

Conversely, when T-bill rates are low or falling, small savers tend to look to markets like stocks, real estate, or bonds to beef up their returns. However, experts contend that falling T-bill rates may show economic weakness, which is not a healthy situation for the market.

Caution: Short-term rates can fluctuate greatly in times of economic uncertainty. Thus, their ability to indicate longer-term trends can be impacted by short-term events.

INTERMARKET ANALYSIS

The domestic and international financial and non-financial markets are interrelated. The stock exchange does not trade by itself; it is directly influenced by the bond market. Bond prices will be affected by the direction of the commodity market. Commodity markets are affected by the trend of the U.S. dollar. The performance of the stock market is often the result of the flows and results in other markets.

Markets do not move by themselves, alone and isolated. In fact, overseas markets affect the U.S. markets, and foreign markets are affected by trends in the U.S. markets. Intermarket analysis applies technical analysis tools to intermarket relationships.

To get an accurate picture of where a particular market is heading, we need to take into consideration all markets that may have an effect on that market. We as responsible technical analysts must realize that no market can be studied alone. Intermarket analysis employs existing technical theory, which has an inward thrust, and moves outward, using information gained from analysis of related markets.

The proper course of action is to use traditional technical analysis market by market, and then expand outward to other markets, taking into account their intermarket relationships. When the data have been collected, the technical analyst needs to see if the individual market conclusions make sense in relation to the intermarket view. If, based on intermarket information, we see an unusual relationship between two markets, we would examine the conclusions derived for the individual markets.

The technical analyst can apply his or her skills to each market separately and without being an expert on any one individual market make intermarket comparisons and develop relationships.

Intermarket analysis requires more information than traditional technical analysis and is therefore harder, but the extra information gained is well worth the effort.

BONDS AND COMMODITY PRICES

The relationship between bonds and commodity prices is one of the most important relationships in the intermarket area. The inverse relationship (see Figure I.3) between Treasury-bond prices and the commodity markets (represented by the Commodity Research Bureau Futures Price Index) shows the connection between the financial sector and the commodity sector that in effect shows a tie between the stock market and the commodity markets. The stock market is greatly affected by the price of Treasury bonds, and stock and bond prices are influenced by the price of the dollar. The U.S. dollar's influence on stocks and bonds finds its origin in the commodity sector. Commodity prices affect bonds, which in turn affect stocks.

Commodity prices are important to watch because they are the major indicator of inflation. In general, when commodity prices rise, we experience inflation; when they fall, inflation is not an issue. Though commodity markets usually move in the opposite direction to bond prices, they move in the same direction as Treasury bond yields. Since the early 1970s every significant turn in long-term interest rates has been paralleled by a shift in the same direction in commodity markets. (We must remember that bond prices and bond yields move in opposing directions.)

Through intermarket analysis we examine these relationships in a little more depth. During a period of inflation, when Treasury bond yields are rising, bond prices will be on the decline. During a period of disinflation, when bond yields are falling, the price of bonds will be rising. The inverse relationship between commodity prices and bond prices can be documented. If we know that bond prices and bond yields move in opposing directions and that commodity prices and interest rate yields head in the same direction, then we can establish that commodity prices and bond prices move in opposite directions. These relationships do exist, and the importance of being technically proficient in analyzing on an intermarket basis is critical for traders and others. In summary, technical analysis of either commodities or bonds would not be complete without a similar technical analysis of the other.

FIGURE I.3—BONDS VERSUS CRB INDEX

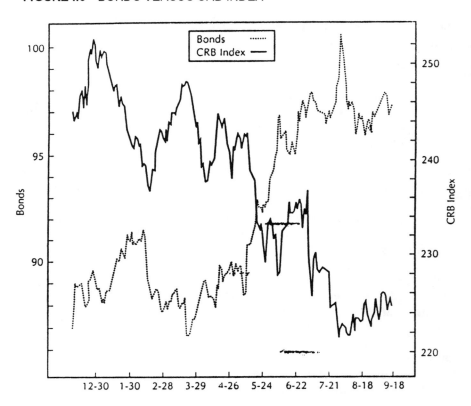

Source: Knight Ridder's Tradecenter. Tradecenter is a registered trademark of Knight Ridder's Financial Information

STOCKS AND BONDS

There is a strong positive relationship between stocks and bonds. The stock market is influenced by interest rates and the way inflation is headed. Usually, falling interest rates are bullish on stocks (prices rising) and rising interest rates make stock prices go down. The technical analyst should be watching the way stocks and bonds behave in relationship to each other. If they are moving together in the same direction, the analyst should not be worried. If they are moving in opposite directions, there should be an investigation as to the reasons why.

Normally, the bond market moves before the stock market. When the stock market is at its peak, the bond market will usually drop first; when the stock market is at its low, the bond market will head upwards first. The conclusion that we can draw from this is that the bond market is a good technical indicator of what the stock market will be doing.

Stocks and bonds have always been tied together by technical analysts and market watchers. An analysis of one without the other would be inadequate. A prudent stock market investor should always be monitoring changes in the bond market. Bond market changes should always be infused into stock market analysis. A technical analysis that is bullish for bonds will also be bullish for stocks and a bearish technical analysis for bonds will be a bearish indicator for stocks.

The business cycle is an important factor in the discussion of stocks and bonds as it relates to periods of expansion and recession. The bond market is usually a strong indicator of the U.S. economy. A weak bond market commonly indicates an economic downturn and a rising bond market indicates an economic surge. Generally, the stock market weakens during an economic drop and benefits from economic expansion.

Although the preceding facts are usually true, there are some differences. A falling bond market is mostly bearish for stocks, but a rising bond market does not always signify a strong equity market. While a rising bond market does not assure a bull market in stocks, a bull market in equities is improbable without a rising bond market.

THE DOLLAR AND COMMODITIES

When analyzing the four sectors of the economy, the path to take is to start with the dollar, move into the commodity market, then the bond market, and finally the stock market. The connection of the dollar to stocks and bonds is more complete and makes more sense when analyzed through the commodity markets.

We can see this if we begin with the stock market and work backwards toward the dollar. The stock market is reactive to interest rates and thus movements in the bond market. The bond market is shaped by expectations of inflation, which are determined by trends in the commodity markets. The inflationary impact of the commodity markets is usually determined by the movement of the U.S. dollar. Thus intermarket analysis should begin with an analysis of the dollar.

A rising dollar has no effect on inflation. A rising dollar usually produces commodity prices that are lower. A fall in commodity prices leads to higher bond prices and lower interest rates. A falling dollar has the converse effect: it is bearish for bonds and equities and bullish for commodities. Thus we can see that the U.S. dollar moves inversely to commodity prices (see Figures I.4 and I.5).

FIGURE I.4—U.S. DOLLAR INDEX VERSUS CRB INDEX: 1985-1989

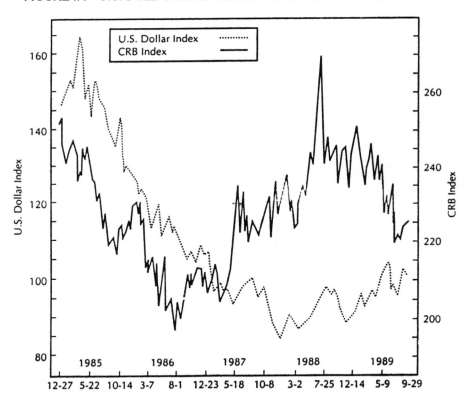

Source: Knight Ridder's Tradecenter. Tradecenter is a registered trademark of Knight Ridder's Financial Information

FIGURE I.5—U.S. DOLLAR INDEX VERSUS CRB INDEX: 1988-1989

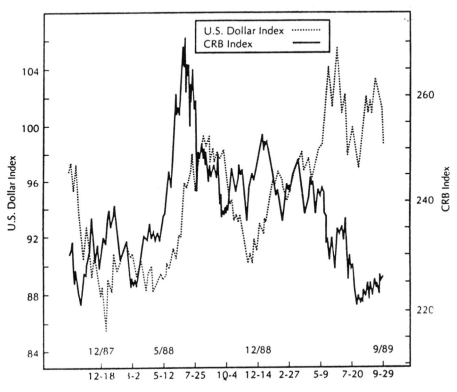

Source: Knight Ridder's Tradecenter. Tradecenter is a registered trademark of Knight Ridder's Financial Information

Figures I.4 and I.5 bring to light some significant facts:

1. A declining dollar is bullish for the CRB index.
2. A rising dollar is bearish for the CRB index.
3. Turns in the dollar happen before turns in the CRB index.

Commodity prices are a leading indicator of inflation, and movements in the U.S. dollar can be used to forecast changes in the CRB index. In the past 28 years every significant turn in the CRB index has been preceded by a turn in the U.S. dollar.

When analyzing these two markets, we need to look at the problem of lead and lag times. We know that turns in the CRB index are preceded by turns in the dollar. If we look at the preceding charts we will see that the peak in the 1985 dollar occurred approximately 18 months before the 1986 fall in the CRB index.

We need a quicker way to predict the impact of the dollar on the commodity markets. Luckily, we can look to the gold market, another step in our intermarket analysis. Of the more than 20 commodity markets that make up the CRB index, gold is the most reactive to changes in the dollar (see Figure I.6). A change in the dollar will quickly produce an opposite change in gold, in turn affecting the general commodity price level.

FIGURE I.6—U.S. DOLLAR INDEX VERSUS GOLD: 1988-1989

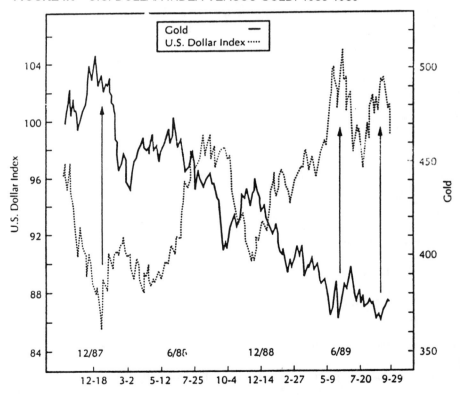

Source: Knight Ridder's Tradecenter. Tradecenter is a registered trademark of Knight Ridder's Financial Information

Major upturns and downturns in the gold market lead those same upturns and downturns in the CRB index. Thus, the lead time between movement in the dollar and the CRB index can be better understood by using the gold market as a link. Gold usually precedes turns in the CRB index by an average of four months.

In summary, there is a direct relationship between the U.S. dollar and the CRB index that will affect stocks and bonds. The falling dollar forces the CRB index higher, and the rising dollar pushes the CRB index lower. Since gold trends in the opposite direction of the U.S. dollar, it establishes a link between the CRB index and the U.S. dollar.

THE DOLLAR, STOCKS, AND INTEREST RATES

A declining dollar ultimately pushes interest rates higher; rising interest rates make the dollar more desirable in relation to other currencies and pulls the dollar higher. The rising dollar eventually moves interest rates lower and lower interest rates make the U.S. dollar less desirable in relation to other currencies, which pulls the dollar lower.

The relationship between the dollar and bonds is easy to see. A declining dollar will push bond prices lower and interest rates higher, and a rising dollar will push bond prices higher and interest rates lower, impacting the stock market.

The U.S. dollar is more reactive to changes in short-term rates than in long-term rates. Long-term rates are based on long-range expectations for inflationary measures. Because short-term rates respond faster to changes in monetary policy, they are more volatile than long-term rates.

We know from our technical analysis that the stock market and the dollar are both affected by interest rate changes, and that there is a positive link between stocks and the dollar, though sometimes with lengthy lead times. A declining dollar will eventually direct the prices of stocks lower due to the rise in interest rates and inflation. A rising dollar will ultimately push interest rates and inflation lower, which will be bullish for stocks.

When we analyze the relationship between stocks and the dollar we must take into account fluctuation in commodity prices (inflation) and changes in the bond market (interest rates). The dollar will affect the stock market only after moving through the other two sectors.

A pattern that forms at market turns aids reversals in stocks, bonds, and the dollar. When interest rates rise, the dollar will turn up first. After a time the advancing dollar will push interest rates lower, and the bond market will see a positive response. Stocks will then move upwards. There will be a period of declining interest rates while bond prices are rising and the U.S. dollar will hit its peak. At this time the dollar will begin to fall and start to push interest rates higher. The bond market will peak and the stock market will follow. This entire process may encompass several years.

In summary, there is a direct relationship between interest rates and the U.S. dollar; they follow each other in a circular fashion. Long-term interest rates affect the dollar less than short-term rates.

All four market sectors—stocks, currencies, interest rates, and commodities—are linked. Intermarket analysis helps us understand the following series of events:

1. The dollar is pushed higher by rising interest rates.
2. The precious metal gold hits its peak.
3. The CRB index hits its peak.
4. Bonds hit bottom and interest rates peak.
5. The stock market bottoms out.
6. The dollar is pushed lower by declining interest rates.
7. The precious metal gold hits bottom.
8. The CRB index hits bottom.
9. Bonds hit their peak and interest rates rise.
10. The stock market peaks.
11. The dollar is pushed higher by rising interest rates (back to step 1)

INTERNATIONAL STOCK MARKETS

The U.S. stock market is one of the three largest markets in the world. The other two are the Japanese and the British markets. These can provide priceless insights into the U.S. market, which is why we should include them in our analysis. For example, a collapse in the Japanese market will affect markets elsewhere.

Global markets usually move in the same direction (see Figure I.7).

The stock markets of all countries do not rise and fall at exactly the same moment or speed, but they are all affected by the general trend of the global market. Bear and bull markets are often experienced on a global basis.

FIGURE I.7—THE THREE MAJOR GLOBAL MARKETS:U.S., JAPAN, AND BRITAIN

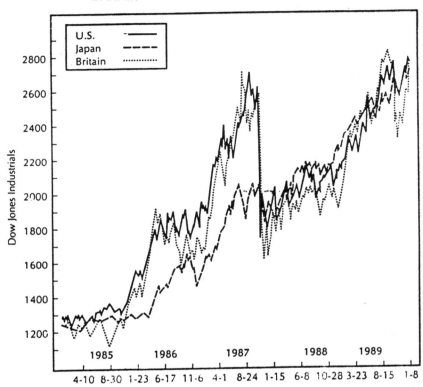

Source: Knight Ridder's Tradecenter. Tradecenter is a registered trademark of Knight Ridder's Financial Information

There is a strong resemblance in the way the U.S. and British markets move, with the UK leading the U.S. market in peaks, as evidenced by Figure I.8.

FIGURE 1.8—U.S. STOCKS (DOW JONES INDUSTRIAL AVERAGE).
BRITISH STOCKS (FT-30 INDEX)

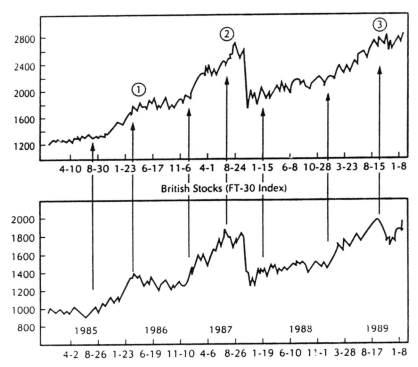

The U.S. and Japanese markets do not fit on such a close trend line as the U.S. and British markets, but it is interesting to note (Figure I.9).
Global markets influence the trend in U.S. stock prices in two ways:

1. U.S. bond prices are influenced by worldwide bond markets.

2. By comparing the U.S. with overseas markets (affected by their own bond markets)

FIGURE 1.9—U.S. STOCKS (DOW JONES INDUSTRIAL AVERAGE)/
JAPANESE STOCKS (NIKKEI 225 INDEX)

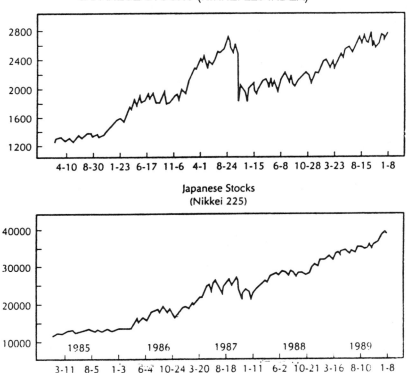

Another important trend to watch is global inflation because inflation trends in the big industrialized countries usually move in the same direction. Though there are some differences in timing, eventually all countries join the worldwide trend. So the prudent technical analyst will watch inflationary trends in other countries as an indicator of potential movement in his own country. This is critical in determining the direction of interest rates that impact on bond and stock market trading and forecasting. World commodity price movement should be watched for clues to where interest rates are headed.

EFFECT OF THE DOW UTILITIES ON STOCKS

The Dow Jones Transportation Average and the Dow Jones Industrial Average form the basis of Dow theory. According to Dow theory, these two averages should trend in the same direction. A third average, the Dow Jones Utility Average, is used to predict changes in the Industrials Average. Utility stocks are closely tied to the stock market through the bond market.

A good technique for the intermarket analyst watching interest-sensitive utilities would be to keep a close eye on the bond market. Utilities usually reflect interest rate changes before those changes are reflected in the stock market. Utilities are also affected by commodity prices and the direction of the dollar. Therefore, no intermarket analysis would be complete without a look at the Dow utilities.

Our intermarket analysis should point out the relationship between the Dow Industrials and the Dow Utilities, which finds its roots in the relationship between bonds and utilities. Usually, the bond market, being very sensitive to inflation, turns first. In addition, utilities that are affected by interest rate changes move in the same direction as bonds but before the general market. The general market turns last and is reflected in the Dow Jones Industrial Average.

The Dow Jones utilities will move in the opposite direction from commodities prices. The Utility Average is an integral part of intermarket analysis as it relates to stocks, bonds, commodity prices and the U.S. dollar. The average should lie between the industrial stock market averages and bonds.

SUMMARY

Generally, the business cycle is an expansion and contraction of the economy, which history has showed us occurs in a complete cycle approximately every four years. The business cycle impacts world financial markets and the relationships between the commodity, stock, bond, and currency markets. We can gain insight into how all four markets interact by studying them in relationship to each other, not by analyzing each one separately. Intermarket technical analysis should take into account all markets, as well as financial and nonfinancial factors, to draw appropriate conclusions. Intermarket analysis considers such factors as the inflation rate, trends in interest rates, the business cycle, economic conditions, and central bank policies.

INVESTMENT STRATEGIES

An investment strategy is a plan to allocate funds among stocks, bonds, commodities, real estate, and other assets. A balanced program is allocating funds equally between long-term and short-term positions in the market. An investor's strategy should take into account his or her tax status, age, risk tolerance, amount of liquid funds, and time horizon, among other factors such as economic factors (e.g., inflation, economic cycle, recession, interest rates) and political concerns.

In general, a *buy and hold* strategy does well in a bull market. Further, the rate of return on a stock often exceeds the inflation rate long term.

Many stocks experience a long-term advance interrupted by minor corrections. For example, a particular stock that has lost much value toward the end of a bear market will be in a solid technical position for an upward advance. The opposite is also true: A company at the top of a bull market may be ready for a long-term decline in price when the downward trend commences. Therefore, life cycles and the characterization of companies must be taken into account in identifying trends in price.

Pyramiding is adding further positions to stock, bonds, or commodities as the market continues in the right direction. The buy or sell signals take place on the first signal.

Some other investment strategies are to:

- Buy the stocks of well-known companies. However, stocks that receive a lot of attention from the mass media may be overpriced.

- Buy leading stocks or groups. Buying a leading stock or group of stocks at the beginning of a bull market usually is quite profitable. Further, there is a tendency in an upward market for certain groups of stocks to outperform the market averages at various stages of a bull market.

- Buy stocks in an out-of-favor industry or company. Out-of-favor stocks may be bought cheap. When their earnings pick up, stock prices will rise. An example is buying interest-sensitive stocks such as banks when interest rates are increasing, if you expect future declining interest rates to benefit financial institutions.

- Invest in undervalued companies.

- Invest in companies that are doing well financially but not being followed by analysts. "Sleepers" have potential to gain in price once they are recognized by Wall Street.

- Fish at the bottom. This may allow you to buy a security at a low price. However, what may appear to be a low price often falls even further, and there may be a valid reason for a steep decline in the price of the stock. A reversal formation at the bottom may reveal that the stock will be on its way up. Is the stock in a turnaround situation in which price has been lowered so much that any good news to come out about the company will cause a drastic price increase?

- Do what the "smart money" is doing, since they have expertise and know best. This includes following insider transactions and tracking program trades.

Example:

Use a top-down approach to investing, looking first at the general economy, then at the state of the overall stock market, then at the industry group, and then at the individual stock. (The same approach can be applied to commodities.) Depending on the time period to be examined, each component should be evaluated from a long-term to intermediate-term to a short-term horizon. Is there a bull or bear phase? In a bull market, there is a greater chance of making money with a particular stock. In a bear market, the likelihood is the stock selected will go down in price.

An investor who believes inflation will remain low may be attracted to buy retail stocks because low inflation improves consumer-spending power.

Note: The converse is the bottom-up approach to investing, in which individual stocks are first reviewed, then the industry, then the overall stock market, and then the economy.

- Select stocks by examining charts for companies emerging from a pulling back to long-term bases.

- Select an industry leader you are optimistic about. If you believe the industry is strong, the industry leader (and for that matter other companies in the industry) will do well.

- Invest based on the January barometer. According to this theory, the performance of the stock market in January predicts what will happen for the whole year. Therefore, if the stock market is up in January, buy stock and if it's down, sell. Related to the January barometer, Yale Hirsch found that if stock prices increase during the first five days in January, the stock market for the entire year will be up—and vice versa: if prices decrease early in January, the market will end the year down.

- Use General Motors as a bellwether stock to predict the performance of the overall stock market. If overall stock market prices are decreasing but the price of General Motors does not go to a new low within four months, then buy. This, too, works in reverse: If GM does not reach a new high, sell.

- Invest based on cyclical economic or industry trends. A cyclical stock can offer much profit opportunity, depending on which business cycle for the industry or economy we are in. The purchase or sale of the stock based on a business cycle analysis is likely to be much more profitable than using a buy and hold strategy.

- Invest based on the contrary opinion principle that the best investment decision is to do the opposite of the general public or Wall Street advisory services. At key market turning points, the general consensus of future direction in security prices is typically incorrect. For example, if the overall consensus of opinion is to sell, the wise decision is to buy.

- Buy on bad news. Shortly after bad news is announced, the price of a company's stock will fall drastically. This may be a good time to buy because there is plenty of room for a rebound, especially if the adverse development is temporary.

A stock may be a buy if stock price and the relative strength line go above key trendlines coupled with a modest increase in volume. In a favorable market environment, buy if price and volume trends support the improving relative strength. A potential gain exists for investors who can select stocks breaking out from extended bases accompanied by increasing volume and an improving relative strength trend long term.

Composite leverage refers to measuring the major variables impacting capital invested.

J

JAPANESE CANDLESTICK CHARTS

Candlestick charts use the same data as bar charts but construct each day's "candlestick" to emphasize the relationship between the day's open and its close. They are so named because the individual chart elements resemble a candlestick with wicks on each end. The thick bar of the candlestick, referred to as the real body, is determined by the day's open and close.

Each candlestick unit represents the trading activity for one time period. Traditionally, in Japan, if the close is higher than the open, the body is colored red; however, this does not copy well so it is common to leave it open (white). When the close is below the open, the body is filled (black). It is also common to refer to white (up) candlesticks as yang and black (down) candlesticks as yin.

The lines that extend from the top and bottom of the real body are called the shadow. They represent the day's high and low trading range.

The close is a period of great activity as day traders and *weak hands* unwind their positions. Weak hands are traders who do not have great confidence in their positions and will easily reverse them. During periods of uncertainty, this will be done before the market closes, as these traders are afraid to suffer losses when they cannot trade out of a position because the markets are closed.

Signals from Japanese candlesticks are more useful in determining trend reversal points than forecasting the magnitude or price objectives of these new trends. Some signals can be generated from a single day's candle; most require two or three days to develop.

Example:

The basic candle patterns are presented in Figure J.1, along with their interpretations. Some patterns change in meaning and name depending on whether the market is up or down.. A hanging man in a market top closely resembles the market bottom signal of a hammer. Other chart patterns closely resemble Western patterns. The three Buddhas chart (not shown) looks like the familiar head and shoulders pattern.

FIGURE J.1—BASIC CANDLE LINES AND PATTERNS

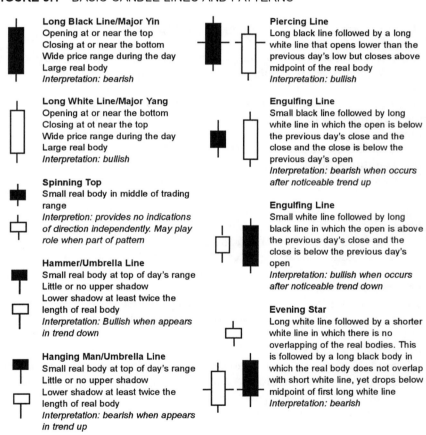

Long Black Line/Major Yin
Opening at or near the top
Closing at or near the bottom
Wide price range during the day
Large real body
Interpretation: bearish

Long White Line/Major Yang
Opening at or near the bottom
Closing at ot near the top
Wide price range during the day
Large real body
Interpretation: bullish

Spinning Top
Small real body in middle of trading
range
*Interpretion: provides no indications
of direction independently. May play
role when part of pattern*

Hammer/Umbrella Line
Small real body at top of day's range
Little or no upper shadow
Lower shadow at least twice the
length of real body
*Interpretation: Bullish when appears
in trend down*

Hanging Man/Umbrella Line
Small real body at top of day's range
Little or no upper shadow
Lower shadow at least twice the
length of real body
*Interpretation: bearish when appears
in trend up*

Piercing Line
Long black line followed by a long
white line that opens lower than the
previous day's low but closes above
midpoint of the real body
Interpretation: bullish

Engulfing Line
Small black line followed by long
white line in which the open is below
the previous day's close and the
close and the close is below the
previous day's open
*Interpretation: bearish when occurs
after noticeable trend up*

Engulfing Line
Small white line followed by long
black line in which the open is above
the previous day's close and the
close is below the previous day's
open
*Interpretation: bullish when occurs
after noticeable trend down*

Evening Star
Long white line followed by a shorter
white line in which there is no
overlapping of the real bodies. This
is followed by a long black body in
which the real body does not overlap
with short white line, yet drops below
midpoint of first long white line
Interpretation: bearish

FIGURE J.1—BASIC CANDLE LINES AND PATTERNS, *con't.*

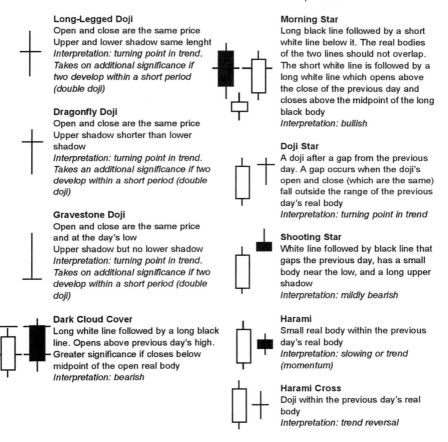

Long-Legged Doji
Open and close are the same price
Upper and lower shadow same lenght
Interpretation: turning point in trend.
Takes on additional significance if
two develop within a short period
(double doji)

Dragonfly Doji
Open and close are the same price
Upper shadow shorter than lower
shadow
Interpretation: turning point in trend.
Takes an additional significance if two
develop within a short period (double
doji)

Gravestone Doji
Open and close are the same price
and at the day's low
Upper shadow but no lower shadow
Interpretation: turning point in trend.
Takes on additional significance if two
develop within a short period (double
doji)

Dark Cloud Cover
Long white line followed by a long black
line. Opens above previous day's high.
Greater significance if closes below
midpoint of the open real body
Interpretation: bearish

Morning Star
Long black line followed by a short
white line below it. The real bodies
of the two lines should not overlap.
The short white line is followed by a
long white line which opens above
the close of the previous day and
closes above the midpoint of the long
black body
Interpretation: bullish

Doji Star
A doji after a gap from the previous
day. A gap occurs when the doji's
open and close (which are the same)
fall outside the range of the previous
day's real body
Interpretation: turning point in trend

Shooting Star
White line followed by black line that
gaps the previous day, has a small
body near the low, and a long upper
shadow
Interpretation: mildly bearish

Harami
Small real body within the previous
day's real body
Interpretation: slowing or trend
(momentum)

Harami Cross
Doji within the previous day's real
body
Interpretation: trend reversal

Computer software for creating Japanese candlestick charts includes:

1. CandlePower (N-Squared Computing, 5318 Forest Ridge Rd., Silverton, OR 97381; (503) 873-4420)

2. MetaStock-Professional (EQUIS International, P.O. Box 26743, Salt Lake City, UT 84126; (800) 882-3040)

3. Compu Trac (Compu Trac Inc., 1017 Pleasant St., New Orleans, LA 70115; (800) 535-7990)

The futures market is well-suited to candlestick charting because the open and close are periods of great activity. Overnight news and rumors help to determine opening price, often resulting in an opening price noticeably different from the previous day's close. In equity markets, the opening price tends to track the previous close unless a dramatic event occurs while the markets are closed.

Candlestick charts can be used for equities, as the Japanese do. Some individuals also recommend candlestick charts in trading options, because certain patterns forecast increases and decreases in market volatility that would affect an options premiums. Options traders could establish positions to benefit from such changes.

L

LEVERAGE

Leverage is the use of fewer funds to make an investment of substantially more value. An example is buying a stock or commodity contract on *margin*.

LIMIT

1. A *limit move* is a price change in a contract that is above the limit established for it on the exchange.
2. *Limit down* or *limit up* is the maximum allowable downward or upward price change for a contract established by a commodity exchange for a particular trading day.

MACD (MOVING AVERAGE CONVERGENCE/ DIVERGENCE)

MACD (pronounced like a Scottish surname, "McDee") stands for Moving Average Convergence/Divergence. The MACD is a trend-following momentum indicator that shows the relationship between two moving averages of prices. Developed by Gerald Appel, publisher of Systems and Forecasts, the MACD is the difference between a 26-day and 12-day exponential moving average. A 9-day exponential moving average, called the "signal" (or "trigger") line, is plotted on the bottom of the MACD to show buy/sell opportunities. (Appel specifies exponential moving averages as percentages. Thus, he refers to these 9-day, 12-day, and 26-day moving averages as 7.5%, 15%, and 20%). The MACD indicator can be applied to the stock market as a whole or to individual stocks or mutual funds.

The MACD proves most effective in wide-swinging trading markets. Three popular ways to use the MACD are crossovers, overbought/oversold conditions, and divergences.

CROSSOVERS

A sell signal (negative breakout) occurs when the histogram moves across the zero line from positive to negative. A buy signal (positive breakout) occurs when the histogram moves above the zero line from negative to positive. The height of the histogram is important, because it reflects the enthusiasm of the market. The greater the enthusiasm, the greater the amplitude of the histogram. When the MACD histogram shrinks and flattens out (i.e., runs parallel to the zero line), the trend is running out of steam, and a reversal is imminent. Movement of the MACD line and stock prices in different directions indicates a price reversal.

OVERBOUGHT/OVERSOLD CONDITIONS

When the shorter moving average pulls away dramatically from the longer moving average (i.e., the MACD rises), it is likely that the security price is overextending and will soon return to more realistic levels. MACD overbought and oversold conditions vary from security to security.

DIVERGENCES

An indication that the current trend may be ending occurs when the MACD diverges from the security. A bearish divergence occurs when the MACD is making new lows while prices are not. A bullish divergence occurs when the MACD is making new highs while prices do not. Both these divergences are most significant when they occur at relatively overbought/oversold levels.

Example 1:

Figure M.1 shows Whirlpool and its MACD. "Buy" arrows were drawn when the MACD rose above its signal line and "sell" arrows when the MACD fell below its signal line. The MACD is calculated by subtracting the value of a 26-day exponential moving average from a 12-day exponential moving average. A 9-day dotted exponential moving average of the MACD (the "signal" line) is then plotted.

Example 2:

In Figure M.2, the three-month chart of the S&P 500, you can see that the black MACD line fell below the red trigger line on January 13, 1999, generating a sell signal. The shaded histogram labeled "divergence" maps the degree to which the MACD line and trigger line diverged. It makes it easier to see exactly where the two have crossed.

FIGURE M.1—MACD FOR WHIRLPOOL CORP.

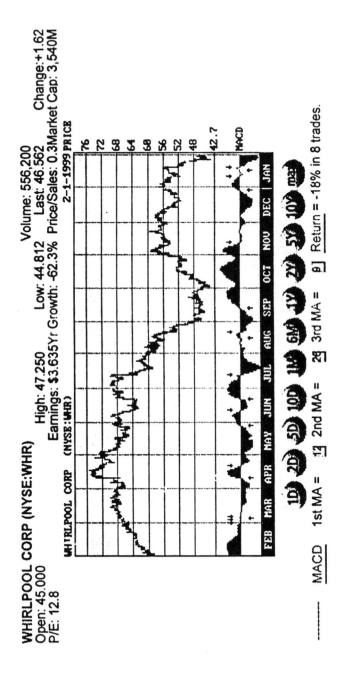

WHIRLPOOL CORP (NYSE:WHR)
Open: 45.000
P/E: 12.8
High: 47.250   Volume: 556,200
Low: 44.812   Last: 46.562   Change:+1.62
Earnings: $3.635   Yr Growth: -62.3%   Price/Sales: 0.3   Market Cap: 3,540M

FIGURE M.2—MACD FOR THE S&P 500 INDEX

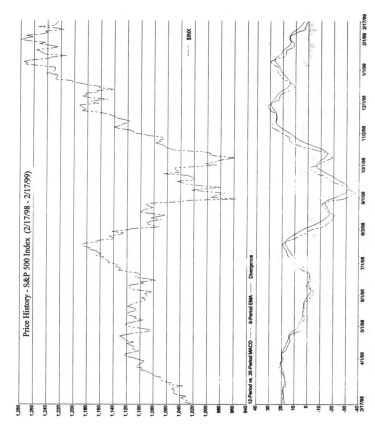

Price History - S&P 500 Index (2/17/98 - 2/17/99)

12-Period vs. 26-Period MACD — — 9-Period EMA —— Divergence

— — $INX

Source: Microsoft Investor
Data Source: SCI

Skeptics may note that the S&P 500 already was down 3% on January 13. But MACD fans would argue that practiced users of the indicator might have been willing to sell a day earlier, escaping all but 1% of the index's decline. Why? Because MACD sell signals are particularly reliable when they follow what the gauge's inventor, Gerald Appel, calls "negative divergences." A negative divergence occurs when a security or index hits a new high but its MACD line fails to do the same. In this example, we see that when the S&P 500 hit a new high on January 8, topping its previous high of November 27, the MACD line failed to reach its own new high.

We see a similar pattern in Figure M.3, the 12-month chart of Chiron (CHIR), a drug stock for which MACD signaled a buy on September 8, 1998, when it was trading at 16 7/16. Immediately after this buy signal, Chiron's stock began a sharp run-up that was interrupted only temporarily in early October when MACD issued a weak sell signal (no negative divergence preceded it). MACD then issued a stronger sell signal in mid-November, when the stock hit a new high but MACD did not, crossing instead under the trigger line. The stock was trading just over $23 that week, up 40% from where it stood when MACD issued its buy signal. Anybody who sold then gave up the final few points of the rally—the stock reached 26 3/8 on January 5, 1999—but escaped the subsequent decline that left it trading at about $21 and trending lower by late January.

MACD also foreshadowed the sell-off in Coca-Cola (KO), when it failed to hit a new intermediate-term closing high in late November 1998 when the stock did. Three trading days later, on November 27, MACD fell below the trigger line, generating a strong sell signal. As Figure M.4 shows, Coke's stock has been in a downtrend ever since, a trend that was not helped by the company's January 26 news that its fourth-quarter profits were down 27% from the previous year. The stock's price action preceded—you could almost say predicted—the news.

MARKET INDICATORS

Market indicators measure an entire market rather than an individual security. They are technical analysis tools to help gauge changes in all securities within a particular market. While market indicators can be used to evaluate any market (stock, bonds, commodities), they are usually used in appraising the stock market.

FIGURE M.3—MACD FOR CHIRON CORP.

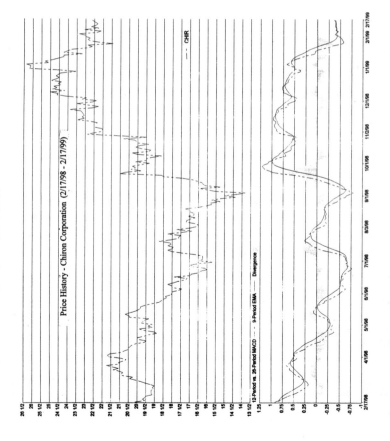

Source: Microsoft Investor
Data Source: SCI

FIGURE M.4—MACD FOR COCA-COLA CO.

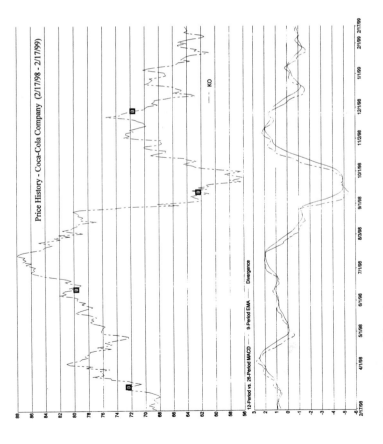

Price History - Coca-Cola Company (2/17/98 - 2/17/99)

—— KO

12-Period vs. 26-Period MACD ---- 9-Period EMA —— Divergence

Source: Microsoft Investor
Data Source: SCI

One common approach to investing is to use market indicators to predict where the market is headed and then use price/volume indicators to ascertain which particular security to buy or sell. The idea is that if the overall market is doing well so should most individual stocks.

The three groups of market indicators are sentiment, momentum, and monetary.

Sentiment indicators look to investor expectations that influence stock prices. The contrarian approach to investing is followed in that the way to market success is to do the opposite of what the majority investors are doing since they are usually wrong. Thus, at a market top, everyone is bullish, so the best investing strategy is to sell; in a market bottom, when investors are gloomy, the best strategy is to buy. For a detailed discussion, see *Sentiment Indicators*.

Momentum indicators reveal what prices are actually doing, taking into account volume of activity. Momentum indicators include comparing volume for stocks increasing or those decreasing in price, advancing versus declining issues, new highs versus new lows, and the price/volume indicators associated with various market indices. See also Momentum and *Momentum Gauges*.

Monetary indicators consider the economic environment such as interest rates, money supply, inflation rate, corporate debt, and consumer debt. A strong economy leads to higher corporate earnings and stock prices.

See also Monetary Indicators.

MARKET RECIPROCAL

The normal average range in volume traded of a stock over a specified number of years divided by the current average range in volume. The higher the result, the less active trading is. A lower result means more active trading.

MEDIAN PRICE

The median price is the midpoint of the daily price. It equals

$$\frac{\text{Low price} + \text{High price}}{2}$$

In Figure M.5 the dotted line represents the average price of a company's stock.

FIGURE M.5—MEDIAN PRICE

Price

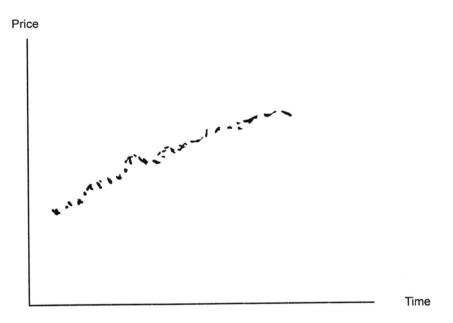

Time

MEMBERS

The participants in a stock exchange who are allowed to purchase or sell securities on it for their own account or those of their clients.

MISERY INDEX

The misery index, developed by Arthur Okun, considers the rate of unemployment and inflation as well as the prime interest rate. The index is inversely related to the broad stock market: It bottoms at market highs and peaks at market lows. Thus, the index is typically negatively correlated to the current condition of the stock market. The misery index has little value as a predictor of future stock prices.

Misery Index = unemployment rate + inflation rate

An alternative way to compute the misery index was developed by Robert Colby and Thomas Meyers is by adding the unemployment rate and the inflation rate to the prime interest rate. The Misery Index may be found in Bureau of Labor Statistics publications and the *Wall Street Journal*.

MOMENTUM

The rate of acceleration of a price or volume movement. Momentum is also the underlying power or thrust behind all upward or downward price movements. Technical analysts examine stock momentum by charting price and volume trends. A stock market advance will be more pronounced if momentum is associated with it.

MOMENTUM GAUGES

Tools used to construct an overbought/oversold oscillator. Momentum measures price differences over a selected span of time. To construct a 10-day momentum line, the closing price 10 days earlier is subtracted from the latest price, The resulting positive or negative value is plotted above or below a zero line.

Momentum is represented on a graph as a line that is continually fluctuating above and below a horizontal equilibrium level that represents the halfway point between extreme readings. Momentum is a generic term embracing many different indicators, such as rate of change (ROC), relative strength indicators (RSIs), and stochastics. Significant gauges of market momentum are explained below.

DOW JONES INDUSTRIAL AVERAGE MOMENTUM RATIO

Measures the percentage difference between the DJIA and its average price for the previous 30 days.

The Dow Jones Industrial Average Momentum Ratio equals:

$$\frac{\text{DJIA closing price}}{\text{DJIA price for preceding 30 days}}$$

A ratio that is more than 3% above the 10-day moving average indicates a market peak. An investor using a contrarian approach would consider it time to sell securities. A ratio that is more than 3% below the 10-day moving average indicates a market trough. A contrarian would buy, with the expectation of increasing prices. This ratio is found in *Barron's*. Figure M.6 presents a momentum chart for Yahoo Inc.

McCLELLAN OSCILLATOR

An oscillator is an internal short- to intermediate-term measure of the strength of momentum. Oscillators use the rate of change to measure a price index. This technical analysis method is used to obtain historical comparative information. Oscillators reflect movements of similar proportion in the same way.

For instance, the *advance/decline line* measures momentum by computing the rate of change in a market average or index over a prescribed time period. It is calculated by subtracting a 39-day exponential moving average of the net difference between the number of advancing issues and the number of declining issues from a 19-day exponential moving average of the net difference between the number of advancing issues and the number of declining issues. New York Stock Exchange data is used in the calculation.

Example:

To construct an index measuring a 26-week rate of change, the current price is divided by the price 25 weeks ago. If the current price is 150 and the price 25 weeks ago was 160, the rate of change (the momentum index) is 93.8 (150/160). The next reading in the index would be determined by dividing next week's price by the price 24 weeks ago.

Oscillators are found in brokerage research reports prepared by technical analysts. They appear, for instance, in the monthly and weekly editions of *Merrill Lynch's Investment Strategy*.

FIGURE M.6—MOMENTUM CHART: YAHOO INC.

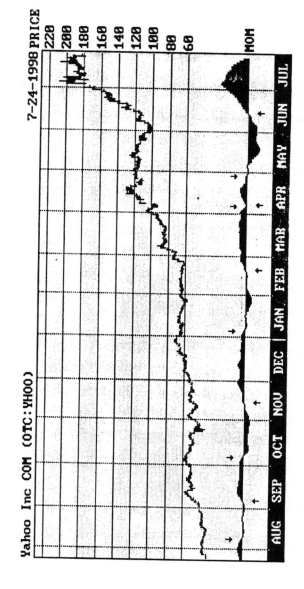

Oscillators can show the rhythm in price movement, which could help the investor determine the degree of momentum associated with stocks. Are stocks in a down or up cycle? Strength in stock price is a bullish indication while weakness is bearish.

The oscillator signals when the market is overbought or oversold. It typically reaches an extreme reading before a change in the trend of stock prices.

The bear market selling climax is indicated by an oscillator reading of approximately -150. A surge of buying activity, implying an overbought condition, is signaled by a reading above 100.

The oscillator normally passes through zero near market tops and bottoms. When the oscillator crosses the zero line going up, it is considered bullish for stock prices. On the other hand, when it goes in the other direction, it is interpreted as bearish.

Numerous chart patterns in the indicator are supposed to signal various market conditions. A full description of how to interpret the McClellan Oscillator is included in *Patterns for Profit: The McClellan Oscillator and Summation Index* (Trade Levels, Inc., 22801 Ventura Boulevard, Suite 210, Woodland Hills, CA 91364).

McCLELLAN SUMMATION INDEX

The McClellan Summation Index is simply a cumulative total of the McClellan Oscillator that is used for interpreting intermediate to long-term moves in the stock market. It is often reported on the cable TV network CNBC.

Like the McClellan Oscillator, the Summation Index gives buy and sell signals. When it crosses zero going up, the signal is bullish; when it crosses going down, the signal is bearish. *Caution:* Reading this index is far from an exact science.

RELATIVE STRENGTH ANALYSIS

Relative strength is the price performance of a company's stock compared to the price performance of an overall market or industry index. In addition, the investor can compare how stocks within an industry have performed relative to the overall market. Market indexes that might be used are the Dow Jones Industrial Average, Standard & Poor's 500, and New York Stock Exchange Composite.

Technicians compute relative strength for a stock using one or both of the following ratios:

$$\frac{\text{Monthly average stock price}}{\text{Monthly average market index}}$$

$$\frac{\text{Monthly average stock price}}{\text{Monthly average industry group index}}$$

An increase in these ratios means that the company's stock is performing better than the overall market or industry. This is a positive sign.

A relative strength index (RSI) also may be computed for a security as follows:

$$\text{RSI} = 100 - \frac{100}{1 + \text{RS}}$$

where

$$\text{RS} = \frac{\text{14-day average of up closing prices}}{\text{14-day average of down closing prices}}$$

A 14-day RSI is mostly used for short-term trading but other periods may be used. A moving average may be used to smooth the RSI.

Relative strength for an industry may be determined as follows:

$$\frac{\text{Specific industry group price index}}{\text{Total market index}}$$

An increasing ratio indicates that the industry is outperforming the market.

Brokerage research reports and *Value Line Investment Survey* may provide relative strength information on companies and industries. Relative strength charts and interpretation may be obtained from Securities Research. You may also compute the monthly average prices of a stock and of a market index by referring to price quotations in newspapers.

Relative strength is an approach that helps the investor determine the quality of an index price trend by comparing it with the trend in another index. An improvement in relative strength after a drastic decline is an indicator of strength. A deterioration in relative strength after a prolonged increase in price signals weakness.

An industry index (e.g., utilities) may historically lead an overall market index (e.g., the DJIA) so that a change in that industry index may imply a future change in the overall market index. A graph may be prepared charting these indexes relative to one another.

Figure M.7 takes a look at the highest 1-year relative strength figures and the total return (price appreciation plus dividends, if any) for actively traded stocks (average daily volume of 2 million or more) for the year ended Oct. 31, 1998:

FIGURE M.7—RELATIVE STRENGTH FOR STOCKS

Company Name	Ticker	Exchange	Total Return 1 Year	Relative Strength 1 Year
Yahoo!	YHOO	Nasdaq	497%	397
Amazon.com	AMZN	Nasdaq	315%	245
America Online	AOL	NYSE	231%	175
Dell Computer	DELL	Nasdaq	227%	172
Lycos	LCOS	Nasdaq	211%	159
Excite	XCIT	Nasdaq	209%	157
Infoseek	SEEK	Nasdaq	179%	132
EMC	EMC	NYSE	130%	91
Apple Computer	AAPL	Nasdaq	118%	81
Wal-Mart Stores	WMT	NYSE	98%	64
Lucent Technologies	LU	NYSE	95%	62
Gateway 2000	GTW	NYSE	92%	60
Ford Motor	F	NYSE	95%	56
Staples	SPLS	Nasdaq	86%	55
Tele-Communications A	TCOMA	Nasdaq	84%	53
Comcast	CMCSK	Nasdaq	80%	49
Ascend Communications	ASND	Nasdaq	79%	49
Novell	NOVL	Nasdaq	76%	47
Cisco Systems	CSCO	Nasdaq	73%	44
Sun Microsystems	SUNW	Nasdaq	70%	42

You can evaluate relative strength to predict individual stock prices, taking a positive view if a stock or industry group outperforms the market as a whole. The presumption is that strong stocks or groups will become even stronger.

A high RSI for a security (stock or bond) or commodity suggests it is overbought, a low RSI that it is oversold.

The 70% and 30% lines in the Wilder RSI graph are used to interpret the graph.

- *70% Line—Overbought Condition:* A top, or overbought condition, is indicated when the RSI goes above 70 percent. The RSI will usually top out before the actual price tops out. When the RSI crosses the 70 percent line going downward, a sell signal (negative breakout) occurs, indicating a new downtrend.

- *30% Line—Oversold Condition:* A bottom, or oversold condition, is indicated when the RSI falls below the 30 percent line. The RSI will usually bottom out before the actual price bottoms out. When the RSI crosses the 30 percent line going upwards, a buy signal (positive breakout) occurs, indicating a new uptrend.

Failure Swings: A second move above 70 or below 30 that is smaller than the previous move is a very strong indicator of a price reversal.

Chart Formations: The index may display chart formations that may not be obvious on the price graph, such as head and shoulder formations, pennants, of flags. Head and shoulder tops or bottoms are one of the most reliable of all major reversal patterns. Pennant and flag formations are true consolidation patterns, indicating a temporary pause in the direction, duration, and target for a major move.

Support and Resistance: Many times, support and resistance will show up on the RSI graph before they appear on the price chart.

Divergence: Divergence between price action and the RSI is a very strong indicator of a market turning point: When the price of the stock continues to rise but the RSI declines, look for a correction.

When the price continues to fall but the RSI rises, look for the stock to rise.

Figure M.8 shows the RSI for Microsoft.

Caution: It is important to distinguish between relative strength in a contracting market and relative strength in an expanding market. When a stock outperforms a major stock average in an advance, it may soon turn around. But when the stock outperforms the rest of the market in a decline, the stock will likely remain strong.

WALL STREET WEEK TECHNICAL MARKET INDEX

This index is based on a survey of 10 technical investment analysis methods. The index is used to substantiate an upward or downward trend in the stock market. Fundamental analysis is not taken into account.

For each of the 10 indicators, this index assigns +1 for a bullish characteristic, -1 for a bearish situation, or 0 for no effect. The ratings are then added to obtain a total. The 10 indicators in the index are:

FIGURE M.8—RSI FOR MICROSOFT, MARCH 1998-FEBRUARY 1999

MICROSOFT CORP (NNM:MSFT) Bid: 147.687   Asked: 147.750 Volume: 18,271,400   Change:+1.93
Open: 147.437   High: 149.250   Low: 145.750   Last: 147.687
P/E: 62.8   Earnings: $2.35 5Yr Growth: 40.7%   Price/Sales: 24.0 Market Cap: 368,267M

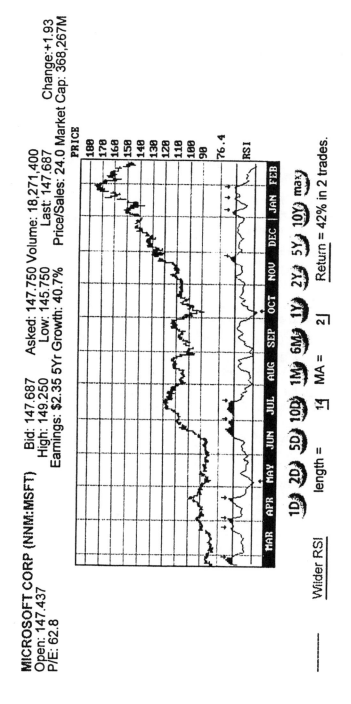

1. *Market Breadth.* A moving average for ten days for the net effect of advancing relative to declining issues.
2. *Put/Call Options Premium.* A ratio of premium on put options to premium on call options.
3. *Arms Short-Term Trading Index.* An advance/decline ratio of NYSE stocks
4. *Insider Buy/Sell Ratio.* A ratio of insider buys to insider sells.
5. *Low-Price Activity Percentage.* The volume of low-priced (risky securities to the volume of high-quality stocks.
6. *Bearish Sentiment Index.* An indicator of how investment newsletters perceive the future condition of the stock market, determined by Investor's Intelligence.
7. *DJIA Momentum Ratio.* The difference between the closing DJIA and the average DJIA for 30 days.
8. *NYSE High-Low Index.* The number of stocks reaching new highs relative to those reaching new lows over the previous 10 trading days, applied on a daily basis.
9. *NYSE Securities at Market Prices above 10-Week and 30-Week Moving Averages.* The percentage of stocks selling above their 10-week and 30-week highs.
10. *Ratio of Ending Prices on Fed Funds to the Discount Rate.* When the Federal Reserve Board tightens the money supply, the Fed rate increases relative to the discount rate because the Fed has set a higher rate between member banks and a lower rate (discount rate) for member banks to borrow from the Fed.

The upper and lower limits for the *Wall Street Week* Technical Market Index is +10 to -10. A reading of +1 to -1 is neutral. A reading of +5 or more is a buy signal; a reading of -5 is a sell signal. The index is found in *Futures* and *Investor's Analysis* magazines.

The index tells from a technical standpoint whether the bottom or top is indicated.

MONETARY INDICATORS

Monetary indicators are based on the theory that interest rates and stock prices are negatively correlated. For example, an increase in interest rates decreases stock prices.

If interest rates drop, the yield on savings accounts and bonds decreases. This makes stocks more attractive because investors will earn a greater return on stocks than on bank accounts or fixed income securities. As interest rates decrease, investors will take their money out of the bank to earn a higher rate of return on stocks. Additionally, lower interest rates results in less corporate borrowing costs, benefiting corporate profitability, which in turn improves stock prices.

The Federal Reserve requirement is the least amount of cash that member banks of the Federal Reserve System must maintain. An increase in reserve requirements is deemed bearish; a decrease in the reserve requirement is bullish, since a relaxation in monetary policy brings an increase in the money supply.

The money supply may act coincidentally with the stock market. An increase in the money supply usually occurs when stock prices are increasing; decreases also track each other. However, the magnitude of changes in the money supply does not indicate how much stocks will increase or decrease.

The *Federal Funds Discount Rate Spread* is the difference between the federal funds rate and the discount rate. An increase or decrease in the discount rate usually comes after a change in the federal funds rate. With a widening spread, there is greater likelihood of an increase or decrease in the discount rate. A positive spread (federal funds rate exceeds the discount rate) means the discount rate will probably be raised. A negative spread has the opposite effect. Weekly data are usually used in the computation.

A discount rate increase is bearish for stocks. Hence, a positive spread (especially more than 3.0) is a bearish indicator. On the other hand, a negative spread (especially below -.3) implies that the discount rate will decrease, which is bullish for the stock market.

The Federal Funds Prime Rate Spread Index is the difference between the federal funds interest rate and the prime interest rate. If the fed funds interest rate increases, the prime interest rate increases. As the spread widens, there is a greater chance that the prime interest rate will be raised or lowered. A positive spread (the fed funds rate exceeds the prime interest rate) means the prime interest rate will probably be raised. A negative spread has the opposite effect.

It is bearish for the stock market if the prime interest rate increases because it will cost companies more to borrow money, lowering their profits. Hence, a positive spread is a negative sign. On the other hand, a decrease in the prime interest rate is good (bullish) for stocks because it reduces a company's borrowing costs, thus improving their profitability. Weekly data are usually used in the computation.

Martin Zweig formulated the *Prime Rate Indicator* as a long-term monetary indicator. The prime interest rate is the rate banks charge their best borrowers. If there is a lag in a change in the prime interest rate compared to other interest rate changes, investors benefit. A change in interest rate comes before changes in stock price. Since a change in the prime rate lags behind other interest rates, it shows when stocks will start to be affected by changes in overall interest rates.

According to Zweig's research the important prime interest rate level is 8%. The Prime Rate Indicator generates two buy and two sell signals, one each for higher than 8% (a high prime rate level) and one each for lower than 8% (a low prime rate level).

The initial buy signal takes place when the prime rate is first lowered from a peak of less than 8%. For instance, if a rate increases in several steps from 4 1/2% to 7% and then goes down to 6 1/2%, this carries a buy recommendation.

At a level higher than 8%, a more substantial change in prime rate is needed for a buy signal. The second buy signal comes on the second of two cuts or a full 1% reduction (which is unlikely) in the prime interest rate from a peak of 8% or more. For instance, if the prime rate increases in several steps to 13% and then is reduced to 12 1/2%, there is no signal yet. An additional reduction from the 12 1/2% level is needed to signal a buy.

The first sell signal is an initial increase in the prime rate from a low of 8% or more. For example, a prime rate decrease from 11% to 9 1/2% in several steps, followed by an increase to 10% is a sell signal.

The second sell signal occurs on the second of two increases or a full 1% increase (very unlikely), in the prime interest rate from a low below 8%. For example, if the prime rate is reduced from 10% to 6 1/2% in several steps and later increased to 7% with another increase to 7 1/2%, there is a sell signal.

Zweig also developed the *Fed Indicator*, a long-term monetary indicator tracking Federal Reserve actions that will give a bullish or bearish sign for stocks. The computation of the Fed Indicator assigns points to the discount rate and Federal reserve requirements. An analysis is then made of the meaningfulness of the score. Positive points are bullish, negative points bearish.

An initial decline in the discount rate or reserve requirements earns two positive points. A subsequent decrease in either the discount rate or reserve requirements earns one positive point. An increase in either component results in one negative point.

An initial positive point will be lost six months after a decrease in the discount rate or reserve requirement, and another initial positive point is lost 12 months after the reduction in the discount rate or reserve requirement. Each subsequent positive point is lost six months after an applicable decrease in the discount rate or reserve requirement. An initial negative point is released six months after it is earned.

When a positive point is earned, all negative points are eliminated, and vice versa.

Once each component has been graded and the two point totals added, the interpretation is:

0 or +1	= neutral
+2 or more	= strongly bullish
-1 or -2	= moderately bearish
-3 or more	= strongly bearish

Norman Fosback has formulated a monetary indicator called *two tumbles and a jump* to forecast increasing stock prices: If the Federal Reserve reduces the reserve requirement, the margin requirement, or the discount rate back-to-back, stock prices will increase.

Edson Gould and Fosback developed a monetary indicator called *three steps and a stumble* to predict falling stock prices: If the Fed increases the reserve requirement, the margin requirement, or the discount rate three times in a row, stock prices will decrease. While this rule may not furnish precise timing it does post a red flag for a major fall in stock prices.

Zweig's *installment debt measure* states that if loan demand is up, interest rates will be up, which is bearish for stocks. A bullish signal comes if loan demand is down. Installment debt on the part of consumers is a measure of loan demand. The installment debt indicator equals:

$$\frac{\text{Current month (e.g., October 20X1) total consumer installment debt}}{\text{Previous year current month (October 20X0) total consumer installment debt.}}$$

Consumer installment debt information is published monthly by the Federal Reserve approximately six weeks after the end of the month.

Zweig has found that a 9% yearly change in consumer installment debt indicates a bearish or bullish environment. If the yearly change is less than 9%, buy; if more, sell.

MONEY FLOW INDEX

Measures the momentum in the market by determining how much money is going into and out of the market. The indicator accounts for volume. To compute the money flow index:

1. Compute the usual price equal to:

   $$\frac{\text{High price} + \text{low price} + \text{closing price}}{3}$$

2. Compute the money flow equal to the usual price multiplied by volume.

3. Determine positive and negative money flow over a specified time period.

4. Compute the money ratio equal to the positive money flow divided by the negative money flow.

5. Calculate the money flow index as follows:

   $$100 - \frac{100}{1 + \text{money ratio}}$$

A decline in the index coupled with increasing security prices signals a forthcoming reversal in the market. If the index is more than 80, a market may have peaked. If the index is less than 20, it may have hit bottom.

MOST ACTIVE ISSUES

Stocks that have the largest share volume for the trading period, usually daily. Active stocks, which reflect the activity of institutional investors, account for about 20% of total NYSE volume. Trading volume for the stocks of each exchange and the over-the-counter market are tabulated.

Statistics on the most active stocks are published in the financial press on both a daily and weekly basis. For example, the *New York Times* lists the 15 most active issues for the trading day on the NYSE, AMEX, and NASDAQ markets, giving volume, last price, and change. *Barron's* lists the high/ low /last prices and change. Prodigy on-line database service lists the most active issues on the NYSE, AMEX, and NASDAQ.

Technical services prepare summaries of the most active stocks. For example, *Indicator Digest* publishes biweekly the trend in up stocks and down stocks.

The quality of the volume leaders will be a key indicator. If they are blue chip companies, the future might be bright. If they are secondary issues, it is wise to be somewhat cautious. The most active issues are highly marketable.

In addition to looking at share volume, consider price movement. If the number of shares traded in a stock is very high but the market price is constant, the security is fairly stable. However, increasing price on heavy volume is significant. The security may be making a significant upward movement. On the other hand, decreasing price on heavy volume may signal a significant downward movement—the "big boys" are unloading, so look out for substantial price changes coupled with heavy trading activity. Is the heavy trading due to an announcement of superior earnings, heavy buying by institutions, or a takeover attempt? Can the volume and price movement be sustained?

The listing of a stock for the first or second time is quite significant and must be thoroughly studied. Some stocks, such as General Electric, AT&T and Philip Morris, are often among the most active because of their wide ownership and institutional backing.

Be alert for repetition. If an industry or a company appears regularly within a short period, something is happening, as evidenced by institutional interest.

Caution: A stock may be very active one day for a special reason such as a possible acquisition. If the rumored acquisition does not materialize, the stock may soon become dormant.

MOVING AVERAGE CONVERGENCE/DIVERGENCE

See MACD.

MOVING AVERAGES

An average that is updated as new investment information is received. The investor uses the most recent stock price, volume, or both to calculate an average that is then used to predict future market prices or volume. Moving averages (MA) can be used to produce buy and sell signals for stock trading.

- When a stock crosses the MA going up, it is a buy signal.
- When a stock crosses the MA going down, it is a sell signal.

The MA is a good trend indicator but often produces whipsaws (i.e., as soon as a signal is given the stock immediately reverses direction and gives the opposite signal). The exponential MA was designed to reduce some of these bad signals, since a single day's large price movement affects it less.

SIMPLE MOVING AVERAGE

The most recent observation is used to calculate a simple MA. Moving averages are constantly updated by averaging a portion of the series and then adding the subsequent number to the numbers already averaged, omitting the first number, and obtaining a new average. For instance, a 30-week MA records the average closing price of a stock for the 30 most recent Fridays. Each week, the total changes because of the addition of the latest week's closing figures and the subtraction of those of 31 weeks ago. The new total is divided by 30 to obtain the MA.

Example 1:

To predict a stock price in June, start with the following month-end stock prices:

January	$ 20	
February	$ 23	
March	$ 20	
April	$ 21	4 months
May	$ 16	

The June price is computed using a moving average as follows:

$$\frac{23 + 20 + 21 + 16}{4} = \frac{80}{4}$$
$$= 20$$

Example 2:

Day	Index	Three-Day Moving Total	Three-Day Moving Average
1	121		
2	130		
3	106	357 (Days 1-3)	119 (357/3)
4	112	348 (Days 2-4)	116 (348/3)

Moving average information and charts may be found in brokerage research reports. Many technical analysts prepare and chart 200-day moving averages. Otherwise, you can determine the MA on a stock from published stock price quotations and volume figures. Typically, daily or weekly price changes are graphed.

Many analysts are of the opinion that a reversal in a significant up trend in the price of a stock or overall market may be identified beforehand, or at least confirmed, by examining the movement of current prices compared to the long-term MA of prices. An MA shows the direction and degree of change of a fluctuating series of prices.

Moving average is used as a model to predict the future expected market price of stock. The investor can choose the number of periods to use on the basis of the relative importance he attaches to old versus current data. For example, an investor might compare a 5-month and a 3-month period, with the data weighted by age. The old data may be given a weight of 4/5 and current data 1/5. In the alternative possibility, the old data is given a weight of 2/3, while current observation gets 1/3 weight.

A 200-day MA of daily ending prices is usually graphed on stock price charts. Buy when the 200-day average line becomes constant or rises after a decline and when the daily price of stock moves up above the average line. A buy also is indicated when the stock price rises above the 200-day line, then goes down toward it but not through it, and then goes up again. Consider selling when the average line becomes constant or slides down after a rise and when the daily stock price goes down through the average line. A sell also is indicated when the stock price is below the average line, rises toward it, but instead of going through it slips down again.

Figures M.9 and M.10 show 10-day and 200-day moving averages for America Online; Figure M.11 presents a 30-day moving average for Microsoft.

FIGURE M.9—DAY MOVING AVERAGE FOR AMERICA ONLINE, INC., JANUARY 15, 1998–JANUARY 15, 1999

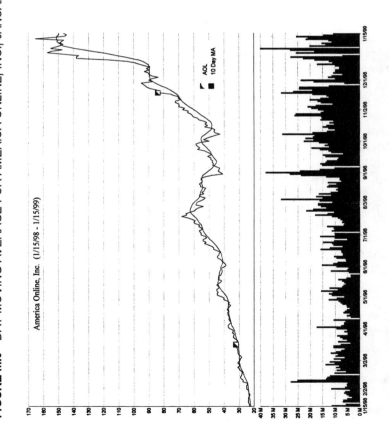

Source: Microsoft Investor

Data Source: SCI

FIGURE M.10—200-DAY MOVING AVERAGE FOR AMERICA ONLINE, INC., JANUARY 15, 1998-JANUARY 15, 1999

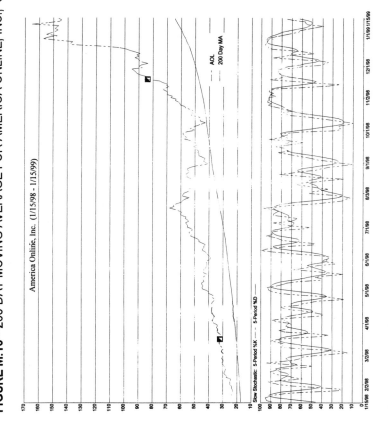

Source: Microsoft Investor
Data Source: SCI

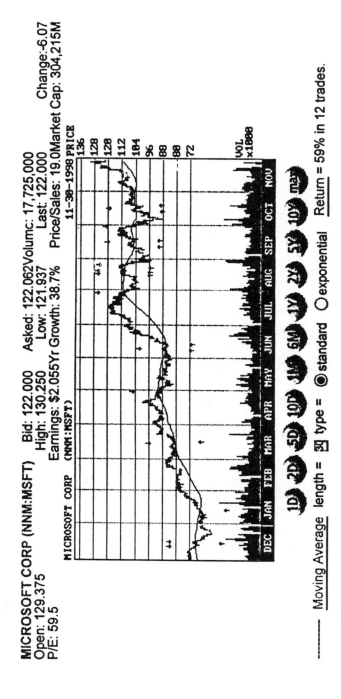

FIGURE M.11—30-DAY MOVING AVERAGE FOR MICROSOFT

FIGURE M.12—30-DAY EXPONENTIAL MOVING AVERAGE FOR DOW JONES INDUSTRIALS 30

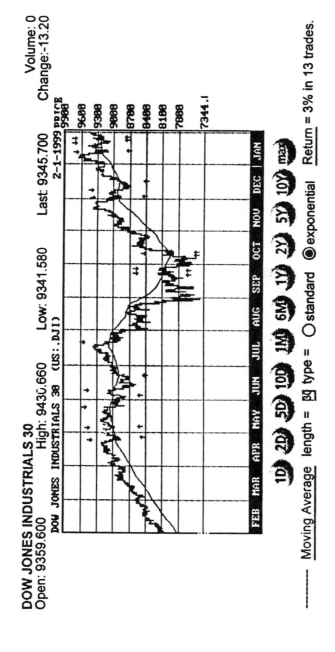

EXPONENTIAL MOVING AVERAGE

To overcome the distortion caused by extreme highs or lows, the exponential moving average weights recent closing prices more heavily than earlier closing prices. Many market technicians consider the exponential MA to be a more accurate indicator than a simple MA.

To calculate an exponential MA, a simple average is first calculated for the desired period. Then the following formula is used for each new moving average:

[Last MA Value * (1 - 2/L+1)] + [NP * 2/L+1]

where:

MA	= Moving average
L	= Length of moving average
NP	= Most recent closing price of stock

Let's say the simple MA of a certain stock over a 19-day period is 100 and the stock closed today at 105. If we plug these figures into the above formula (Last MA Value = 100, L = 19, and NP (New Price) = 105), the new MA will be 100.5. The same formula is used to figure consecutive values for the remaining periods.

Figure M.12 on page 195 shows a 30-day exponential moving average for Dow Jones Industrials 30.

MOVING AVERAGES IN COMMERCIAL CHART SERVICES

Several companies publish charts on stocks that show price fluctuations, volume of trading, and a moving average. Among them are:

Company Name and Address	Chart series	Length of Moving Averages
William O'Neilll P.O. Box 66919 Los Angeles, CA 90099-5925	Daily graphs	50 and 200 days
R.W. Mansfield 2973 Kennedy Blvd. Jersey City, NJ 07306	Weekly charts N.Y., AMEX, or over-the-counter stocks	50 and 150 days

Company Name and Address	Chart series	Length of Moving Averages
Securities Research 101 Prescott St. Wellesley Hills, MA 02181-3319	21-month security charts	65 and 195 days
Trendline Division of Standard & Poor's 25 Broadway New York, NY 10004	Daily action	50 and 150 days
	Current market perspectives	150 days

N

NEGATIVE VOLUME INDEX (NVI)

The negative volume index (NVI) is based on the theory that trading by unsophisticated investors occurs predominantly on days of high volume, while sophisticated investors trade during quieter periods. Thus, a negative change in volume reflects buying and selling of stocks by those in the know.

Usually, stock prices drop whenever the volume drops. However, when the stock price rises on lower volume days, it is considered very positive. NVI is calculated by adding the stock's percentage gain or loss to the previous value whenever the volume drops from the preceding period.

Table N.1 shows the calculated performance of the S&P 500 when its NVI was in an uptrend. As can be seen, a strong NVI can be very bullish.

TABLE N.1—S&P'S PERFORMANCE AND NVI TREND

Trend of NVI	Probability Bull Market Is in Progress	Probability Bear Market Is in Progress
Up	96%	4%
Down	47%	53%

If the NVI is trending upward, the stock is likely to continue to rise. If the NVI is trending downward, it is ambiguous. Figure N.1 shows AOL and its NVI.

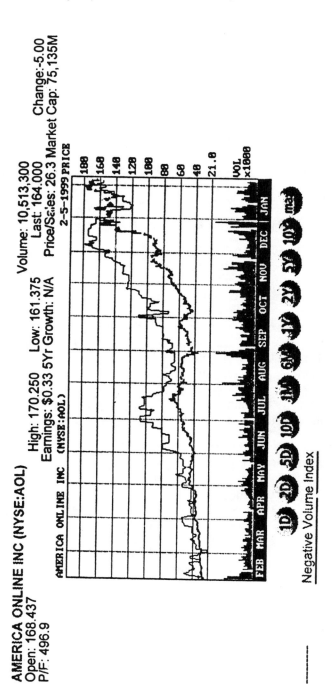

FIGURE N.1—NEGATIVE VOLUME INDEX FOR AMERICA ONLINE

NEW HIGHS AND NEW LOWS

New highs are those stocks for the trading day that hit their 52-week high;
new lows are those that hit a 52-week low. They are listed as "highs" and
"lows" in daily financial newspapers such as the Wall Street Journal and
Investors Business Daily. An index higher than 1 indicates that the number of stocks reaching new highs are more than those making new lows.
A ratio above 10 is bullish; a low ratio (below .06) is bearish. Graphs
might be prepared as shown in Figure N.2. Daily or weekly New York
Stock Exchange (NYSE) data are usually used in the computation.

FIGURE N.2—NEW HIGHS VERSUS NEW LOWS

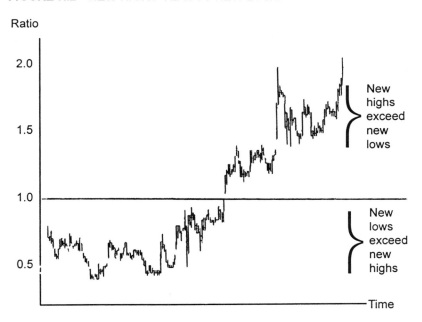

Technicians view the relationship of new highs to new lows as an
indicator of stock market trends and momentum. The number of new
highs to the number of new lows is a very low ratio just before a princi-
pal market bottom. This may signal a buying opportunity. On the other

hand, a very high ratio of new highs to new lows occurs at market ceilings. This may signal selling—especially for contrarians.

If the overall market indices are reaching new highs but fewer and fewer stocks are reaching new highs, this implies a weak uptrend, a bearish divergence that may indicate a future downward trend in prices.

A 15-day moving average of new highs-new lows may be plotted to smooth daily figures.

A positive ratio of new highs to new lows indicates a bullish condition; a negative ratio is bearish. Thus, an uptrend in the ratio of new highs to new lows is a bullish indicator for the stock market while a downtrend is bearish.

The new highs/new lows ratio equals:

<u>Number of new highs</u>
Number of new lows

Example: Two issues have hit new 52-week highs and 10 have hit new lows. The ratio is therefore .20, which is low. This is a bearish situation.

In general, an increasing trend of new highs to new lows is consistent with increases in major market indices. However, if the new highs/new lows indicator goes in the opposite direction to market indices, the divergence suggests a possible reversal in the overall market. For example, if new highs/new lows sharply drop while the market indices are increasing, the overall market will probably decline.

A high/low logic index was formulated by Norman Fosback. Using daily or weekly NYSE information, it equals:

<u>Lower of number of new highs or new lows</u>
Total number of issues traded

The theory behind this approach is that many stocks will be going to new highs or to new lows but typically not both at the same time.

A high reading of the high low logic index indicates inconsistency because many stocks are at new highs while many stocks are at new lows. This is a bearish indicator. A very low ratio indicates consistency and uniformity, a bullish sign. Table N.2 provides guidelines for the high/low logic index:

TABLE N.2—HIGH/LOW LOGIC INDEX GUIDELINES

Time Period	Reading	Indication
Weekly	In excess of .05	Bearish
Weekly	Below .01	Bullish
Daily	In excess of .018	Bearish
Daily	Below .003	Bullish

A 10-period simple moving average (MA) may be used based on daily or weekly figures to smooth out unusual movements.

The technician should evaluate the number of stocks making new lows (on a 52-week basis) for a particular week or day. A 10-period MA of daily data is recommended. A very low reading (e.g., 4 or less) is a very bullish indicator.

If in a major market averages after a long advance with consecutive higher peaks, the new number of new highs reflect a series of declining peaks, a red flag is posted. This scenario implies a gradual weakening from a technical perspective, since successive peaks in the market averages are associated with fewer stocks hitting new highs.

The net number of new highs also considers stocks experiencing new lows. In a bear market, a new low in a stock index (e.g., the S&P 500) that is not coupled with a decreasing number of net new highs is a positive indicator. In such a situation, a decreasing number of new lows implies fewer downside breakouts.

The investor may look at specific companies having new highs as the basis to make an investment decision. With a stock that appears to be overvalued, the investor should not buy it or, if it is currently held, sell it. Of course, if the investor believes the company will do even better, the stock should be retained. On the other hand, a company listed as a new low might be bought if the investor feels the price is overly depressed and the company has potential.

NIKKEI GOLF MEMBERSHIP INDEX

In Japan, business executives like to play golf. This weekly index tracks the average price of golf membership in the 400 leading clubs in Japan against a base of 100 set up on January 1, 1982. Before 1985, the index was a lagging indicator of stock prices on the Tokyo Stock Exchange. After the stock crash of 1985, memberships were sold to meet margin calls and cover stock market losses. The index gives signals about the stock market in Japan.

ON-BALANCE VOLUME TECHNICAL INDICATOR

Developed by Joseph Granville, the indicator calculates on-balance volume by adding volume to the previous value when the stock price closes higher than the previous price, and subtracting volume from the value when the stock closes lower than the previous price.

If the stock falls to a new low and the on-balance volume line does not, the indicator is positive (bullish divergence). If the stock rises to a new high and the on-balance volume line does not, the interpretation is negative (bearish divergence).

OPEN INTEREST

The number of options or futures contracts still open when the trading day ends. An open contract is one that has not expired, been exercised, or been closed out. Open interest figures are mined by commodity traders for demand and supply information.

Open interest affects both the buyer and seller of the futures contract. It increases for the parties when a new contract arises, whether because the buyer has a long position or the seller a short position. If contracts are closed because the buyer sells or the seller covers his short position, the open interest decreases.

Open interest reflects the liquidity associated with a particular contract or market because a reduction or increase in open interest indicates that funds are going out of or into an option or futures contract. Further, an increase in both volume and open interest confirm the existing trend. However, if volume and open interest decrease together, the current trend will end. Thus, an appraisal of the mix of volume and open interest suggests how money is moving into and out of the market.

Open interest is the number of contracts trading for a specific commodity. Here is how the implications of open interest can be interpreted:

Open Interest	Price	Conclusion
Up	Down	Bearish
Up	Up	Bullish
Down	Up	Bearish
Down	Down	Bullish

An increasing open interest in an uptrend indicates aggressive new buying which is bullish; in a downtrend, it indicates aggressive short selling, which is bearish—as is decreasing open interest in an uptrend. On the other hand, a decreasing open interest in a downtrend infers the selling of unprofitable long positions. This is bullish.

When an increasing price trend is ending, a decline or flatness in open interest signals that a reversal in price trend is imminent.

A selling climax occurs after a long decline in price. When prices suddenly decline sharply on heavy volume, there is a significant decline in open interest.

OSCILLATORS

Oscillators, which are short- to intermediate-term breadth indicators, are an internal measure of the strength of momentum. They measure a price index by looking at the rate of change (ROC). This technical analysis method is used to obtain historical comparative information. Oscillators reflect movements of similar proportion in the same way. An example of a momentum oscillator is the advance/decline line.

Momentum is measured, using NYSE data, by computing the rate of change in a market average or index over a prescribed period by subtracting a 39-day exponential moving average of the net difference between the number of advancing issues and the number of declining issues from a 19-day exponential moving average of the net difference between the number of advancing issues and the number of declining issues.

Example: To construct an index measuring a 26-week rate of change, the current price is divided by the price 25 weeks ago. If the current price is 150 and the price 25 weeks ago was 160, the rate of change (momentum index) is 93.8 (150/160). The next reading in the index would be determined by dividing next week's price by the price 24 weeks ago.

Oscillators are found in brokerage research reports prepared by technical analysts and in the monthly and weekly editions of Merrill Lynch's *Investment Strategy*.

Oscillators can show the rhythm in price movement, which could help the investor determine the degree of momentum associated with stocks. Are stocks in a down or up cycle? Strength in stock price is a bullish indication while weakness in stock prices is a bearish sign.

The oscillator signals when the market is overbought or oversold. It typically reaches an extreme reading previous to a change in the trend of stock prices.

A bear market selling climax is indicated by an oscillator reading of about -150. A surge of buying activity, implying an overbought condition, is signaled by an oscillator reading above 100.

The oscillator normally passes through zero near market tops and bottoms. When the oscillator crosses the zero line heading up, it is considered bullish. On the other hand, when it crosses the zero line going down, it is interpreted as bearish.

See also Breadth Analysis, especially Advance/Decline Line; Momentum Gauges, especially McClellan Oscillator.

OVERBOUGHT/OVERSOLD INDICATORS

If the overall market or a particular stock is overbought (overvalued), it is vulnerable to a downward correction, so the prudent thing to do is sell. In an overbought market the prices of securities may have increased very steeply and too fast. In other words, there has been aggressive buying so the overpriced stocks are very vulnerable to a decline in price. On the other hand, an oversold (undervalued) overall market or particular stock would prompt the investor to buy because of an expected upturn in prices. In an oversold market, selling of securities has been excessive. Danger zones are zones of price movement where the market or security is extremely overbought or oversold.

Some technical indicators measure whether the market or a specific stock is overbought or oversold. One measure is the +30%/-10% rule as related to the Standard and Poor's (S&P) 500 Index: If the 12-month change of the S&P 500 exceeds 30%, an overbought market is indicated, providing a sell signal. If the 12-month change of the S&P 500 is below

10%, an oversold condition is indicated, signaling a buy. Another measure of an overbought/oversold market looks at the last 10 trading days. The indicator equals:

$$\frac{(\text{Current price}) - (\text{10-day low price})}{(\text{10 day high price}) - (\text{10-day low price})}$$

The market is overbought if the indicator exceeds 90 and oversold if the indicator is below 10.

OVERHANG

A significant block of securities or commodities contracts that if sold would result in downward pressure on prices or if bought would cause upward pressure.

PARABOLIC

1. The parabolic time/price system formulated by Welles Wilder is used to set trailing price stops. It is a stop and reversal approach signaling appropriate exit points. When the price increases above the stop-reversal point, a short position should be covered. When price goes below the stop-reversal point, on the other hand, a long position should be closed
2. A parabolic curve occurs when the increase in stock price accelerates to the point that it is increasing in practically a straight line. This will last for only a short time and usually results in a long decline from the peak price.

PAY UP

A pay up occurs when an investor desiring to own a stock at a specified price hesitates and the price starts to increase. Rather than letting the stock go, the investor "pays up," buying the shares at the higher prevailing price.

PENETRATION

Penetration occurs when a stock price breaks through one side of a trendline or boundary to the other side.

PENNANT

A horizontal chart pattern characterized by two converging trendlines. A pennant resembles a pointed flag, with the point facing to the right (see Figure P.1). It is preceded by a significant and almost straight line direction resulting when a substantial advance or decline gets before itself, with the market pausing before taking off again in the same direction. A pennant is formed as the rallies and peaks that provide its shape become less pronounced. The pennant is characterized by declining trade volume. Once the pattern is finalized, prices will increase or decline significantly.

FIGURE P.1—PENNANT

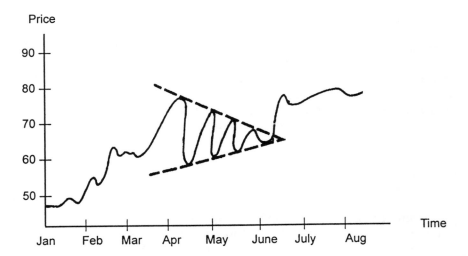

PERFORMANCE CHART

A performance chart shows the percentage change in the price of a stock. The base is the initial period shown on the chart. An indicator of 8, for example, signifies that the stock price has risen 8% from the first period. However, an indicator of -5 means the price has decreased 5% from the first period. The calculation of the performance indicator is:

$$\frac{\text{Current close - initial close}}{\text{Initial close}} \quad \text{x } 100$$

A performance chart (see Figure P.2) is useful in comparing price changes of different stocks.

FIGURE P.2—PERFORMANCE CHART

PLATFORM

Occasionally, after the midpoint of a saucer bottom, there is an abrupt upward price on heavy volume. After that, prices resume a slow rounding process. A platform sometimes occurs toward the end of the base, followed by a resuming new uptrend.

PORTFOLIO THEORY

A sophisticated investment approach in which the investor classifies, estimates, and controls return and risk. The risk-return tradeoff is emphasized, as are the correlations among individual securities comprising the overall portfolio after a determination of how much should be invested in different classes of investments. The stress is on portfolio optimization.

POSITIVE VOLUME INDEX (PVI) TECHNICAL INDICATOR

The positive volume index (PVI) focuses on market action during periods of increasing volume. It is calculated as the opposite of the *negative volume index (NVI)*, with the percentage change in stock price used only during periods when volume is rising. A declining PVI indicates decreasing prices on rising volume, which is considered negative. Figure P.3 shows Microsoft and its PVI.

PRECIOUS METALS

Precious metals—particularly gold—arouse the feelings of investors like no other market or commodity. Since gold stimulates the emotions of investors, the price of gold moves in trends. There are a vast number of analytical tools available to the technical analyst with regard to gold, thus making it easier to analyze than most other commodities.

One of the most popular technical analysis tools is trend analysis. Trend analysis can be applied to the relationship between gold and gold shares, gold and other precious metals, the price of gold itself, and the relationship between gold and its advance/decline (A/D) line.

The price of gold is influenced by three things:

1. Gold as an emotional barometer

2. Gold used as a metal in industry

3. Gold used as money

Gold is a finite substance, unlike any other currency in the world that can be increased at a moment's notice. The emotional nature of gold and its pricing is most evident in the investor's predictions regarding the

FIGURE P.3—POSITIVE VALUE INDEX (PVI) FOR MICROSOFT

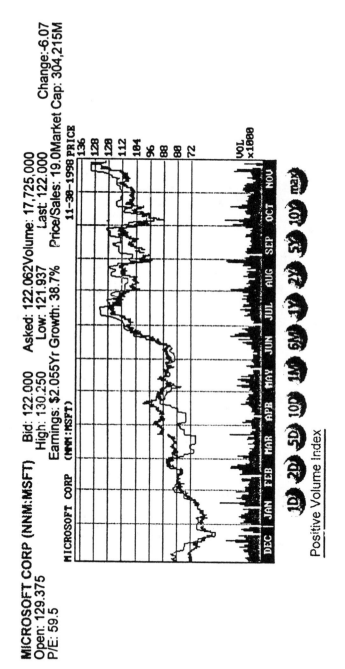

MICROSOFT CORP (NNM:MSFT)   Bid: 122.000   Asked: 122.062   Volume: 17,725,000
Open: 129.375   High: 130.250   Low: 121.937   Last: 122.000   Change:-6.07
P/E: 59.5   Earnings: $2.055   Yr Growth: 38.7%   Price/Sales: 19.0   Market Cap: 304,215M

Positive Volume Index

future course of inflation. Occasionally, the U.S. dollar becomes weak in the world market; this is the time to buy gold as a reserve asset.

The price of gold is so determined by emotion that even political events or speculation about future political events will make the price of gold fluctuate. Though this emotion is in no way related to the business cycle, it increases a rally in a bull market and lessens declines in a bear market.

THE PRICE OF GOLD

Trend analysis techniques that are used to analyze the price of gold itself are moving averages, price pattern behavior, and rates of change (see Figure P.4).

FIGURE P.4—GOLD BULLION MONTHLY

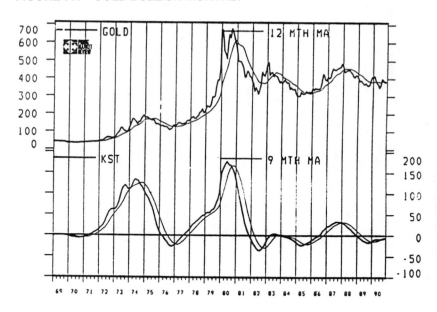

Source: Pring Market Review

THE GOLD ADVANCE/DECLINE LINE

The advance/decline line for gold is measured by taking the net number of advances or declines and then adding or subtracting that number from the running total. The weekly price of gold is measured in 12 different currencies; the gold A/D line can be formed from any number of these (see Figures P.5 and P.6).

The whole idea underlying the formation of the gold A/D line is to estimate whether the price rise or decline of gold bullion is wide-ranging or fluctuating in response to movements in the price of the U.S. dollar.

FIGURE P.5—GOLD *VERSUS* THE GOLD A/D LINE: 1974-1993

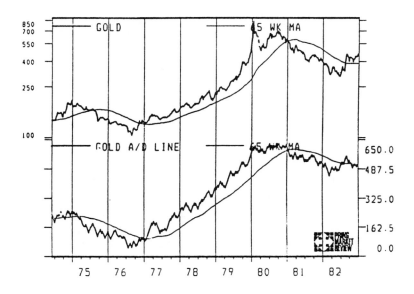

Source: Pring Market Review

FIGURE P. 6—THE GOLD A/D LINE: 1982-1990

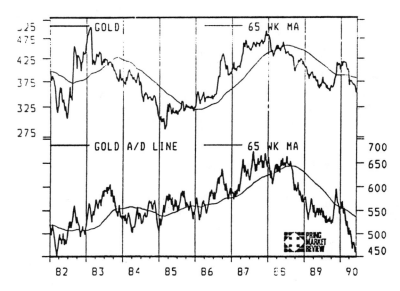

Source: Pring Market Review

GOLD AND GOLD SHARES

The price of gold bullion is often determined by gold shares. There are many different factors that can distort the price of gold mining companies based on information that is totally unrelated to the price of gold. It is therefore most important to consider the various gold share indices, such as the Philadelphia Gold and Silver Index (XAU) or the Toronto Stock Exchange Gold and Silver Share Index (TSE), when analyzing the price of gold.

Investors are interested in the profit margin of gold shares which is led by the average price of gold bullion over a given period. When investors speculate that the average price will fall, they will usually sell in advance. Generally, the current trend is assumed to remain in force unless a new high or low in bullion or stocks is not confirmed. This could be taken as a warning that the present trend is likely to reverse (see Figure P.7).

FIGURE P.7—THE GOLD PRICE VERSUS GOLD SHARES (IN U.S.
DOLLARS), MONTHLY

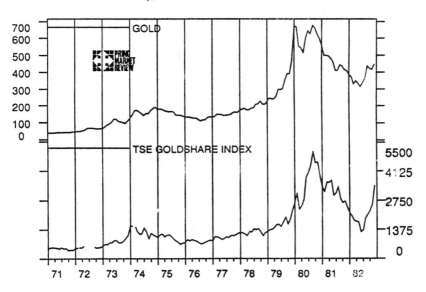

Source: Pring Market Review

GOLD AND OTHER PRECIOUS METALS

During the normal business cycle, usually the four precious metals—silver, palladium, gold, and platinum—move in the same direction. The
most important markets to watch are platinum, gold, and silver. If two of
these three markets drop, a trend reversal warning should be apparent.

Relative Strength plays an important part in predicting trend reversal
as it relates to the price of gold (see Figure P.8).

The gold market has a tremendous inverse relationship to the U.S.
dollar, and the direction of the gold market is a key part in the perceived
direction of inflation.

FIGURE P.8—GOLD VERSUS SILVER, 1974-1983

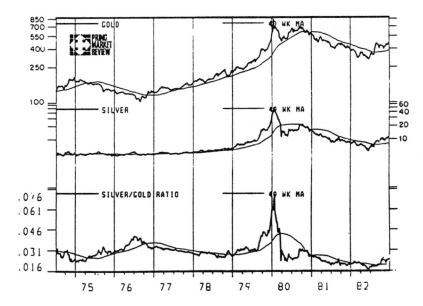

Source: Pring Market Review

PRICE CALCULATIONS

TYPICAL PRICE

The average daily price. It equals:

$$\frac{\text{Close} + \text{Low} + \text{High}}{3}$$

WEIGHTED CLOSE PRICE

The average daily price with more weight on closing price. The weighted close is computed as follows:

$$\frac{(\text{Close} \times 2) + \text{Low} + \text{High}}{4}$$

PRICE AND VOLUME TREND (PVT)

The price and volume trend (PVT) represents cumulative volume adjusted for changes in closing prices. The PVT adds or subtracts part of daily volume. The volume added to PVT is based on how much prices increased or decreased compared to the previous day's close.

The PVT shows how much money is being put into or taken out of a security. It adds a small part of the volume to the indicator if there is a minor change in price but a large portion if there is a substantial percentage change in price. PVT equals:

<(Today's close - yesterday's close/yesterday's close) x volume> + yesterday's PVT

Example: A stock price went up by 1% on volume of 20,000 shares. Therefore, we add 200 (0.01 x 20,000) to the PVT.

PRICE PATTERNS

Illustrations or groups of data that are common on price charts of stocks or commodities, A study of price patterns uses historical experience to predict future prices.

Among common patterns are distribution and accumulation formations, continuation patterns, area patterns, ascending and descending triangles, and reversal patterns.

A *distribution formation* is a pattern formed at market tops that indicates that the stock or market may be experiencing a distribution from strong knowledgeable investors to weak uninformed buyers.

An *accumulation formation* is a price pattern forming at market bottoms that indicates that the stock or market may be in the process of accumulation from weak naïve investors to sophisticated investors.

A *continuation pattern* is a price formation constituting temporary interruptions of the prevailing trend. It points to the continuation of an existing trend. A continuation pattern may include a symmetrical triangle (coil) noting a pause in the existing trend, after which the initial trend is resumed.

An *area pattern* is one in which a stock or commodity's upward or downward momentum is temporarily exhausted, resulting in a flat, horizontal—sideways—price movement.

An *ascending triangle* is a pattern categorized as a flat upper trendline with the lower line rising. The pattern shows more aggressive buying than selling, a bullish indicator.

A *descending triangle* is a pattern evidencing more aggressive selling than buying. The downside signal is indicated by a decisive close below the lower trendline, typically with higher volume.

A moving average should be used to smooth price patterns so as to allow the investor to better understand price trends.

A *spike* top or bottom generally is a non-pattern.

A *whipsaw* is a false signal in trend.

Reversal patterns are important enough to deserve extensive discussion.

REVERSAL PATTERNS

Reversals indicate that the trend is changing. In analyzing reversal patterns, the technical analyst will study the volume patterns that accompany the price information to see if the reversal pattern can be relied on.

A reversal pattern (see Figure P.9) is an area pattern that breaks out in a direction opposite the previous trend. A bottom reversal pattern is one in which prices hold above a previous low; in a top reversal, prices hold below a previous high. There is confirmation when the previous peak of a bottom or trough of a top is overcome.

The following points apply to most reversal patterns:

1. As the pattern increases in size, the ensuing move becomes greater.

2. As we see upward movement, volume becomes more important.

3. Small price ranges are usually found on the bottom and grow slowly.

4. A break in a critical trendline should signal a trend reversal.

5. Bottom patterns last longer than top patterns and are less subject to change.

6. A reversal pattern can only appear if there is a trend.

7. A reversal price pattern typically comes before a reversal of the current major trend.

There are many different types of reversal patterns, some of the more important ones will be explained next.

A *key reversal day* (Figure P. 10) is evident on a bar chart when the daily price spread on the reversal day consumes the previous day's spread, and closes outside it. The signal given is more crucial than the reversal day itself. The markets are usually erratic on heavy volume on the key reversal day. In a downtrend, prices open up lower but close higher for the day. In an uptrend, prices open up higher but close lower for the day. A reversal has a high probability when the price range is wider on the key reversal day.

FIGURE P.9—REVERSAL PATTERN

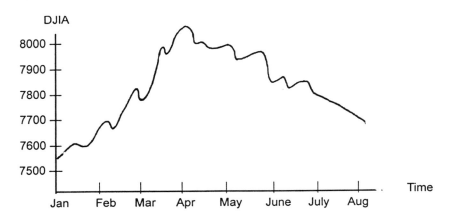

FIGURE P.10—PRICE STRUCTURE OF KEY REVERSAL DAY

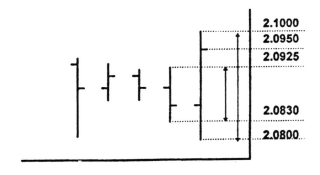

Key reversals are found not only in currency futures bar charts but also in the cash markets.

Head and shoulders reversal patterns are the most widely recognized reversal patterns (see Figures P.11A and P.11B). They occur when a major uptrend begins to lose its drive in its peaks and valley. The uptrend will level off while supply and demand are equal. Upon the completion of the distribution phase, levels of support noticeable along the bottom of the trading range are shattered and establishment of a new downtrend has occurred.

Volume is an important factor in the formation of the head and shoulders pattern. Lighter volume is usually associated with the head (second peak) than with the left shoulder. This usually shows a move towards decreased buying pressure. Volume should be substantially lighter during the third peak (right shoulder). As can be seen, volume is less critical during the culmination of market tops.

FIGURE P.11A—HEAD AND SHOULDERS REVERSAL PATTERN

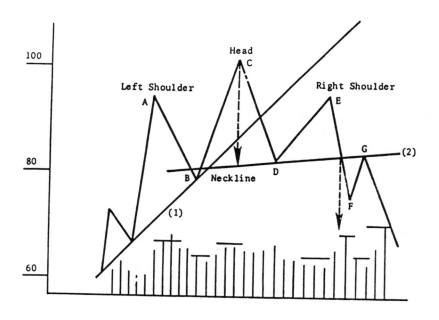

FIGURE P.11B—HEAD AND SHOULDERS REVERSAL PATTERN

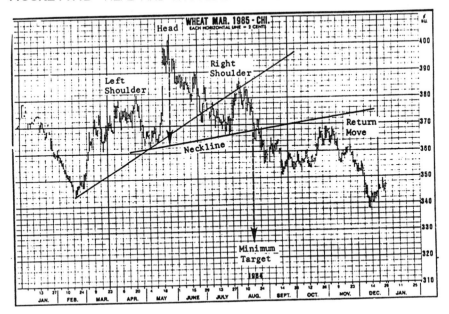

The *inverse head and shoulders reversal* (see Figure P.12) is also called the head and shoulders bottom and usually has the same characteristics as the head and shoulders top except that it is inversed. Normally, there are three definitive bottoms with the middle section protruding lower than the two shoulders. Usually, an abrupt close near the neckline will be required to finish the pattern. A minor difference occurs at the bottom after a bullish market turnaround and causes a tendency for movement back towards the neckline.

There are differences between the top and bottom head and shoulders patterns, the most important one being the sequence of volume. Volume is crucial to the fulfillment and classification of the head and shoulders bottom. In general, when bottom patterns are concerned, markets usually exhibit a marked increase in buying. In response, volume will increase and initiate a fresh bull market.

FIGURE P.12—INVERSE HEAD AND SHOULDERS REVERSAL

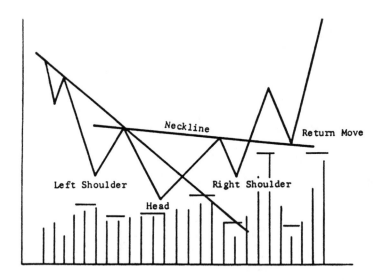

Complex head and shoulders reversal patterns are a variation of the usual head and shoulders pattern that does not occur often. The pattern exhibits a double left or right shoulder as well as a double head formation. There is noticeable symmetry in the head and shoulders pattern.

The *triple tops and bottoms reversal pattern* is a small and rare variation of the head and shoulders pattern. The difference is noticeable in that the three peaks are at the same level. Often, technical analysts are in disagreement as to whether the reversal is a triple top or a head and shoulders formation. It does not matter because both patterns have the same implications.

Figures P.13A and P.13B show a triple top and a triple bottom.

After the head and shoulders pattern, the *double top or bottom reversal patterns* (Figures P.14A and P.14B) are the most common. They are easily distinguishable: There are only two peaks, whether on the top or bottom, and again, volume is important.

Figures P.14A and P.14B depict a double top and bottom reversal pattern.

FIGURE P.13A—TRIPLE TOP

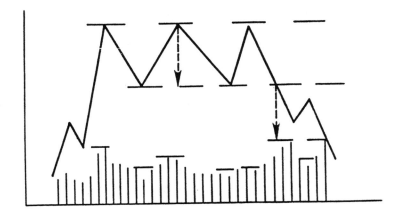

FIGURE P.13B—TRIPLE BOTTOM

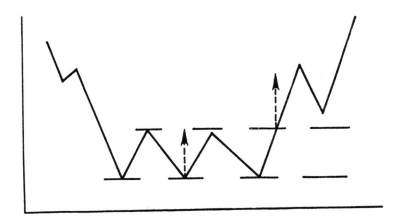

FIGURE P.14A—DOUBLE TOP REVERSAL PATTERN

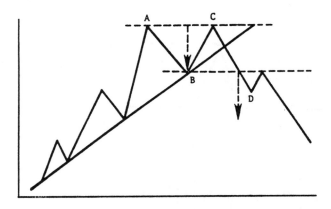

FIGURE 14.B—DOUBLE BOTTOM REVERSAL PATTERN

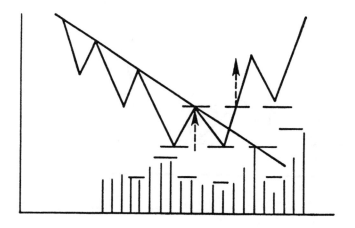

V FORMATIONS, OR SPIKES

A V formation is a technical chart pattern forming a V (Figures P.15A, P.15B, and P.15C). A V pattern reflects the fact that the stock, bond, or commodity has bottomed out and is now in a bullish trend. However, an inverse V pattern shows the opposite and is bearish.

This common reversal pattern is difficult to distinguish. It is different from the usual tendency for a market to change direction. There is usually no warning before the sudden change of trend and a rapid move in the opposite direction. Usually these changes are initiated by key reversal days.

FIGURES P.15A & B—V FORMATIONS

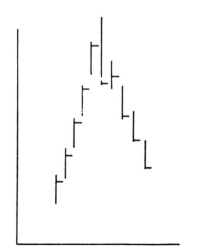
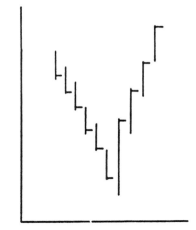

CONCLUSION

If there is a reversal in trend direction, price usually shows a reversal pattern. The pattern is more important the larger and deeper it is.

FIGURE P.15C—V FORMATION

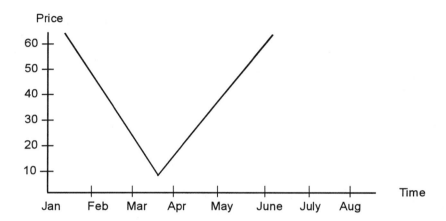

PROGRAM TRADING

The simultaneous purchase or sale of at least 15 different stocks with a total value of $1 million or more. A tabulation is made of the number of shares transacted for the day and published in the Wall Street Journal based on information furnished by the New York Stock Exchange.

PULLBACK OR THROWBACK

1. *Pullback* occurs when the overall market or a particular security falls back in price from a previous advance.
2. *Throwback* is the return of the price of a particular stock or overall market to the downside after an upside breakout.

PUTS/CALLS RATIO

Developed by Martin Zweig, the Puts/Calls Ratio (P/C Ratio) is a market sentiment indicator that shows the relationship between the number of puts to calls traded on the Chicago Board Options Exchange (CBOE). A put is the right to sell a stock at a fixed price by a specified date. A call is the right to buy a stock at a fixed price by a certain date. The P/C ratio is calculated by dividing the volume of puts by the volume of calls:

Total CBOE Put Volume
Total CBOE Call Volume

The P/C ratio increases when there is more put activity due to pessimism around a market bottom; it decreases due to more call activity from investor optimism around a market peak.

The option buy (initial option transaction establishing a long position) call percentage looks at open buy call transactions to total call volume. Investor optimism is reflected in a high ratio, while a low ratio indicates investor caution. Traditionally, options are traded by unsophisticated, impatient investors who are lured by the potential for huge profits with a small capital outlay. Interestingly, the actions of these investors provide excellent signals for market tops and bottoms.

Because investors who buy calls expect the market to rise and those who buy puts expect the market to decline, the relationship between the number of puts and calls illustrates the bullish/bearish expectations of these traditionally ineffective investors.

The higher the P/C ratio, the more bearish these investors are. Conversely, lower readings indicate high call volume and thus bullish expectations. The P/C ratio is a contrarian indicator. When it reaches "excessive" levels, the market usually corrects by moving in the opposite direction. Table P.1 gives general guidelines for interpreting the P/C ratio. However, the market does not have to correct itself just because investors are excessive in their beliefs! As with all technical analysis tools, use the P/C ratio in conjunction with other market indicators.

TABLE P.1

	P/C Ratio 10-day moving average	P/C Ratio 4-week moving average
Excessively bearish (buy)	Greater than 80	Greater than 70
Excessively bullish sell)	Less than 45	Less than 40

Example: Figure P.16 shows the S&P 500 and a four-week moving average of the Puts/Calls Ratio. "Buy" arrows show when investors were excessively pessimistic (greater than 70) and "sell" arrows when they were excessively optimistic (less than 40). The arrows certainly show that investors are buying puts when they should be buying calls, and vice versa.

FIGURE P.16—FOUR-WEEK MOVING AVERAGE OF THE PUTS/CALLS RATIO FOR THE S&P 500

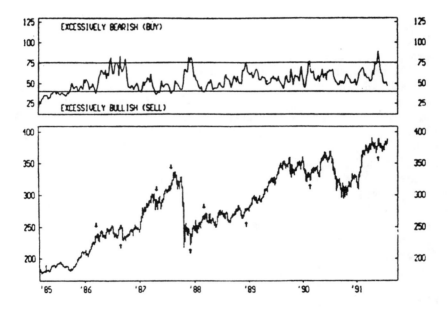

PYRAMIDING

Holding profits from a securities or commodities position as collateral to purchase additional positions with funds borrowed from a broker. The use of leverage creates increased profits in a bull market. However, in a bear market there is potential for larger losses arising from margin calls.

QUADRANT LINES

A series of horizontal lines dividing the range between the lowest and the highest prices (or volume) into four equal parts. They aid in following price changes because they show the lowest, highest, and average price for a particular period.

The quadrant lines are based on the lowest low and highest high for a stated period. The bottom line is charted at the lowest price with the top line drawn at the highest. The other three lines are drawn to divide the area between the lowest low and highest high into four equal parts. The mean is depicted by the center, drawn as a dotted line.

R

RALLY

An increase in the prices of stocks or commodities futures generally during a general declining trend in market prices. The rally may arise from investor bargain-hunting or because a support level has been reached. Unfortunately, technical rallies do not last long; prices then continue their downward pattern.

RANDOM WALK

The random walk theory presumes that price movements of stocks or commodities take place without a pattern, i.e., that present security prices are independent of previous ones. The market price of a stock moves randomly around real (intrinsic) value, but there are periodic changes in the intrinsic value arising from new information. New data affecting stock price are immediately reflected in market value. They enter the marketplace in a random way.

Random walk is based on the efficient market hypothesis. In consequence, random walk supports a buy-and-hold strategy rather than timing of buys and sells. According to random walk, price history is not an accurate barometer of future price direction. The theory basically says that an investor who throws darts to pick stocks has a good chance to outperform the market.

Technical analysts strongly disagree with the random walk theory. Technicians argue that charts of previous price changes allow for the reliable prediction of future price movements.

RANGE

The low and high price points of a stock, bond, or commodity over a stated period. An example is the 52-week high and low price figures for stocks. It is significant from a technical perspective when stock price breaks out of its trading range, whether on the upside or downside. For example, a stock moving out of the upper boundary may go higher.

REACTION

A short-lived period of declining security prices occurring after an upswing. It may arise from profit taking or negative developments.

RELATIVE COMPARATIVE STRENGTH

A comparison of the performance of a particular company's stock with that of major competitors, industry averages, and the overall market.

RELATIVE STRENGTH ANALYSIS

A basic analytical concept in the arsenal of the technical decision-maker, relative strength analysis compares the performance of a financial instrument such as a stock with that of an industry group as an aid in stock selection. The goal is to invest in stocks with improving relative strength. The method is to calculate if the stock price is increasing or decreasing more rapidly than the group price. In other words, technical analysis will determine if the first item is performing better or worse performance than the second.

Relative strength analysis is usually used to contrast an industry group's performance with that of the broad-based market or a certain security's performance to that of its industry group. For example, the following ratio is enlightening:

$$\frac{\text{Monthly average stock price}}{\text{Monthly average market (or industry group) index}}$$

An increase in this ratio means the stock is doing better than the overall market or industry.

In a broader sense, relative strength analysis can be used to contrast any two items that have their own prices, such as industry groups, commodities, or stocks.

At the heart of relative strength analysis is the Relative Strength Index developed by Welles Wilder Jr. in his 1978 book *New Concepts in Technical Trading Systems*. The index was developed as a system to give actual buy and sell signals in a dynamic market. Relative strength indicators are significantly more volatile than price.

Relative strength (RS) is an extremely important concept. Of the many ways to measure RS, the most common is to compare an individual item within a market to a base, as when General Motors is compared to the Standard & Poor's Composite. RS is a useful tool for the selection of individual stocks and securities or in deciding whether to buy, say, gold or bonds. It can be used to compare foreign currencies.

Suggestion: A top-down investment approach to relative strength is to:

Step 1. Ascertain which industry groups are doing best in the broad market.

Step 2. Determine the best performing stocks in those industries.

Step 3. Use technical analysis techniques such as chart patterns to determine when to buy the stocks selected in step 2. (Step 3 deals with timing.)

Increasing RS of a stock compared to the industry is desirable. The stock should be bought in a price uptrend. Conversely, stocks with low relative strengths compared to the industry groups or overall stock market should be sold short. If both a stock index (e.g., S&P Airlines) and the RS line break above their down trendlines, this may represent a strong buy signal for a stock group. Declining RS after a long-term uptrend indicates weakness; the converse—increasing RS after a downtrend equals strength—is also true.

A leading group often shows an improving trend in RS at the start of a cycle. If it does not, it is inconsistent because it should be the strongest at this point, and you have technical weakness.

RS may be plotted weekly, monthly, etc. A moving average of RS may also be plotted. RS is often used to determine price tops and bottoms by focusing on specific levels (usually 30 and 70) on the Relative Strength chart, which has a range of 0 to 100.

The Relative Strength Index can also be used to indicate:

1. Things that do not normally appear on a bar chart.

2. Warnings of impending reversals and swing failures higher than 70 and lower than 30. In *bottom failure swing* the RS index moves in a downtrend (under 30), does not make a new low, and then moves up past its previous peak.

3. Impending reversal based on noticeable differences between the price and the Relative Strength Index.

4. Support and resistance levels appearing with increased clarity.

5. Divergence between price and RS, an early indicator of a problem that will later be confirmed by a trend reversal signal.

The Relative Strength Index needs lead time to be considered accurate. A change in RS may be spotted around key intermediate points in the market.

Normally, the most important trend pattern monitored by all technical analysts is divergence. Divergence is the change in pattern of a trend line following a pattern flowing in the same general direction and indicating a weakness in the trend in the present direction. Usually this type of movement signals a reversal in the trend that can be relied on with a high degree of confidence. Market traders normally wait until a divergence is confirmed before initiating any transactions so as not to be fooled by erroneous signals. (See Figure R.1.) RS analysis may also confirm an existing trend.

The 14-day Relative Strength Index is the most popular measurement, though other time frames can be used. The 6-day index is popular for intraday analysis to determine profit points after the Relative Strength Index has reached 75 or above and is pulling back. The shorter RS Index is highly responsive in pointing out a pause in the current trend. There is also the even more sensitive 3-day Relative Strength Index for determining the most advantageous reentry point as the indicator adjusts itself between 30 and 70, showing neither overbuying nor overselling.

A popular Relative Strength Index divergence to be used along with financial statements is the 10-week model. It often gives excellent signals in terms of the indices of bonds, currencies and stocks (see Figure R.2).

Investment reports (e.g., *Value Line Investment Survey*) provide relative strength data on particular companies, as do financial newspapers such as *Investor's Business Daily*, which publishes relative strength figures for thousands of individual companies and 200 industry groups. Relative strength charts and interpretations may be obtained from Securities Research.

FIGURE R.1—TREND CONFIRMATION: TREASURY BONDS

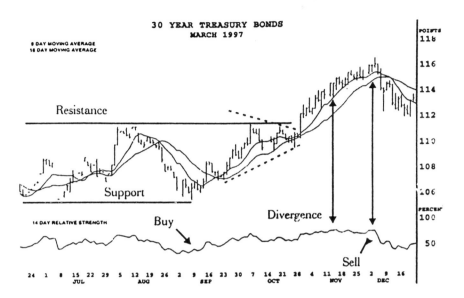

FIGURE R.2—TEN-WEEK MODEL

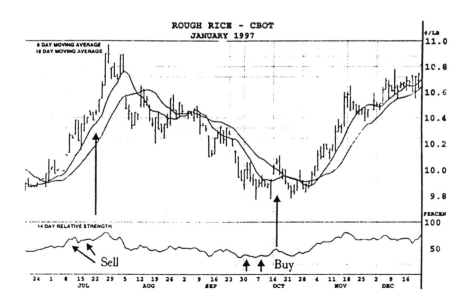

Investment reports (e.g., Value Line Investment Survey) provide relative strength data on particular companies, as do financial newspapers such as Investor's Business Daily, which publishes relative strength figures for thousands of individual companies and 200 industry groups. Relative strength charts and interpretations may be obtained from Securities Research.

RELATIVE STRENGTH INDEX CALCULATION

The Relative Strength Index is calculated as follows:

$$RSI = 100 - \frac{100}{1 + RS}$$

$$RS = \frac{\text{Average of N period's up closes}}{\text{Average of N period's down closes}}$$

N   = number of periods used in the calculation

Using 14 days, the most popular period, as our example, to arrive at the average up value add the total points gained on the up days during the previous 14-day period and divide by 14. The average down value is calculated similarly for the total points lost on down days. The next step is to divide the average up value by the average down value to determine the RS. Insert the RS calculation into the denominator of the formula to calculate the initial day's Relative Strength Index. To recalculate RS each period, multiply the previous up and down average values by 13, add the latest day's gain or loss, and divide the total by 14. Adjust the Relative Strength Index formula by inserting the new RS value and recalculate the index. Table R.1 shows the calculation of RS for 14 days.

Example: Assume that you are interested in the relative strength of a stock for the period of one month that has 20 exchange days. The trading data follow (U = an increase and D = a decrease).

Day	1	2	3	4	5	6	7	8	9	10
Price	10U	3D	4D	5D	2U	4D	6U	1D	6D	7U

Day	11	12	13	14	15	16	17	18	19	20
Price	8U	3D	5U	6D	1D	3D	4U	1D	3U	2U

$$RS = \frac{52U/10}{32D/10} = \frac{5.2}{3.2} = 1.625$$

TABLE R.1—EXAMPLE OF HOW TO CALCULATE THE RELATIVE STRENGTH INDEX

Day	Closing Price	Change in Price from Previous Day		Totals for Last 14 Days		Averages for Last 14 Days		Up Avg. Divided by Down Avg.				RSI
		Up	Down	Up (A)	Down (B)	Up (A)/14=(C)	Down (B)/14=(D)	(C)/(D)=(E)	1+(E)=(F)	100/(F)=(G)	100-(G)	
1	43.000											
2	44.125	1.125										
3	43.250		0.875									
4	42.875		0.375									
5	43.000	0.125										
6	42.875		0.125									
7	42.625		0.250									
8	42.125		0.500									
9	42.750	0.625										
10	43.750	1.000										
11	44.125	0.375										
12	43.750		0.375									
13	44.500	0.750										
14	44.125		0.375	4.000	2.875	0.286	0.205	1.391	2.391	41.818	58.182	
15	44.125			4.000	2.875	0.286	0.205	1.391	2.391	41.818	58.182	
16	44.875	0.750		3.625	2.875	0.259	0.205	1.261	2.261	44.231	55.769	
17	45.000	0.125		3.750	2.000	0.268	0.143	1.875	2.875	34.783	65.217	
18	44.375		0.625	3.750	2.250	0.268	0.161	1.667	2.667	37.500	62.500	
19	43.875		0.500	3.625	2.750	0.259	0.196	1.318	2.318	43.137	56.863	
20	42.750		1.125	3.625	3.750	0.259	0.268	0.967	1.967	50.847	49.153	
21	42.625		0.125	3.625	3.625	0.259	0.259	1.000	2.000	50.000	50.000	
22	44.000	1.375		5.000	3.125	0.357	0.223	1.600	2.600	38.462	61.538	
23	43.750		0.250	4.375	3.375	0.313	0.241	1.296	2.296	43.548	56.452	

RS may be used in the futures markets by looking at price spreads between different products.

Generally, the Relative Strength Index is subject to more volatility, distortions, and unpredictable movement due to its use of ratios than are smoothed indicators, which do not rely on ratios. There are a greater number of signals and therefore transaction costs with this type of measurement than with smoothed momentum indicators.

Technical analysts keep an eye on market trends. They do not buy stocks based on high RS or sell stocks because they have low RS. They analyze the overall market trend, during an uptrend buying high relative strength stocks and during a downtrend sell low relative strength stocks.

Figure R.3 presents RS and a trendline violation while Figure R.4 shows RS and a price pattern.

FIGURE R.3—RS AND TRENDLINE VIOLATION

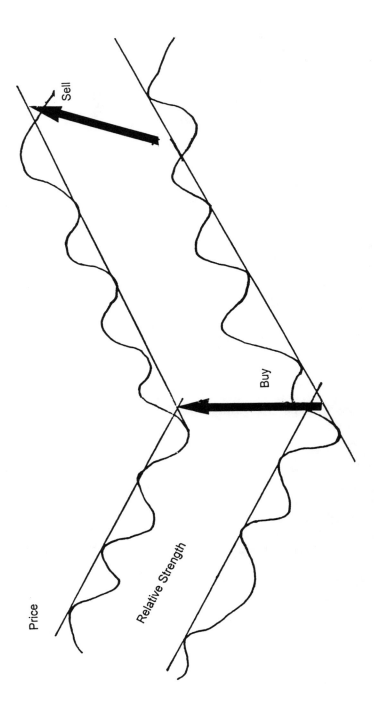

FIGURE R.4—RS AND PRICE PATTERN

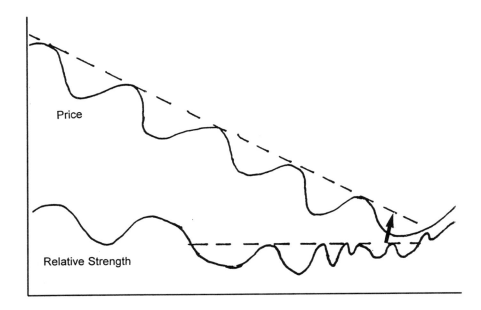

A 15-day Relative Strength Index (plotted at the bottom of the chart in Figure R.5) is moving upward along with XYZ Co.'s stock price. However, its move above the 60 line implies that the stock may be approaching an overbought level, signaling caution.

CONCLUSION

The various Relative Strength Indexes are useful as counter-trend oscillators. The indexes are rate-of-change oscillators measuring the momentum or velocity of price movement. They point to overbought (high RSI value) and oversold (low RSI value) conditions in the stock, bond, and commodities markets. Relative strength analysis aids in identifying major tops and bottoms. Evaluating RS assists in predicting individual stock and industry group prices.

One premise supported by some is that strong stocks or groups of stock will get even stronger, while weak ones will get weaker. On the other hand, some technicians postulate that a stock group significantly advancing will tire out and reverse itself. As a general rule, a stock that does better than the rest of the market in a decline will usually remain strong.

FIGURE R.5—15-DAY RS INDEX

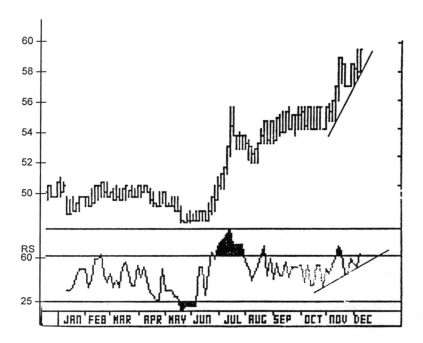

It should be noted that RS often peaks ahead of price, and its doing so is a positive sign in a declining market. RS and price usually move in the same direction. A firming of the RS line in a downward price movement reflects growing technical strength. However, stocks that do poorly relative to the market should be avoided or sold short.

RS analysis may be applied to compare the performance of two different financial assets, such as stocks and bonds, or of items within a similar asset class, such as silver and gold.

Relative strength for a stock may be computed relative to the market or industry group as follows:

$$\frac{\text{Monthly average stock price}}{\text{Monthly average market (or industry group) index}}$$

An increase in the ratio means the stock price is performing better than the overall market or industry.

RS usually runs strong in a long-term trend. Stocks or groups of stock in a strong upward movement in relative strength terms will probably continue to do better than the overall market until the momentum subsides and significantly alters direction. This may take a number of months. On the other hand, stocks or groups with highly negative RS can do worse than the overall market for an extended period.

RESISTANCE

See Support and Resistance.

RETRACEMENT

Repeating some of all of a previous movement, such as a movement in stock price or volume. Percentage retracement is when prices retrace part of a previous trend before continuing in the original direction. An example is a one-third retracement. Countertrend moves tend to go in predictable percentage parameters. In most cases, an action that retraces more than two-thirds of the trend in process signals a reversal. It is not viewed as a temporary situation.

In practical terms, an indication of a strong bull market is that it is going to higher highs and higher lows. This is evidence of improvement in investor expectations and demand. However, retracement looks at the fall back in these prices. Percent retracement looks at how much prices retreat after a higher high, based on the percentage prices retreated from the high to the low. This measure aids in ascertaining when prices will reverse and then continue upward.

A major retracement of 65% strongly suggests a bull market is over. A minor retracement (e.g., 10%) sometimes occurs over the course of a continuous bull market.

Figure R.6 of XYZ Company may prove enlightening. Point 1 is the price before a price move, point 2 is the price at the end of the move, and point 3 is the retraced price.

FIGURE R.6—RETRACEMENT

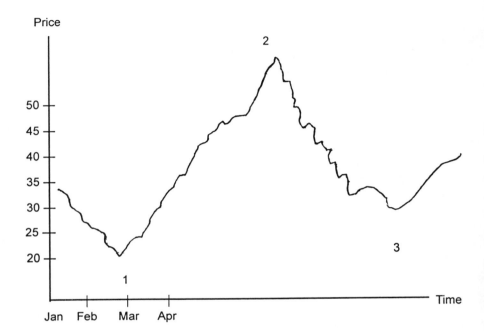

Example: Assume stock prices went from a low of 100 to a high of 150 and then moved back to 125. The move from 150 to 125 (25 points) retraced half (25/50) of the move from 100 to 150 (50 points).

REVERSAL DAY

A trading day on which prices establish a new extreme in the direction of the trend. Prices close in the opposite direction.

A two-day reversal is one in which if an uptrend exists, prices make a new high for a move and end the trading day near that high. However, in the next trading day prices do not move higher but instead open unchanged and then close near the previous day's low. An uptrend is a series of gradually ascending peaks and troughs; a downtrend a similar series but a descending one. The two-day reversal in a downtrend is a mirror image, with first new lows, an unchanged open, and then a close near the previous high.

A *key reversal day* is a turning point but is not correctly seen as such until after the fact. In other words, it occurs after prices have moved substantially in the opposite direction of the previous trend.

The *outside day* is the day when the high and low on a reversal day exceed the range of the previous day.

REVERSAL FORMATION

A pattern where the direction of price movement is changing from up to down or down to up. Stock price approaches the pattern from one direction but exits in the opposite direction.

RISING BOTTOMS

A pattern depicting an increasing trend in the low prices of a stock, bond, or commodity (see Figure R.7). Rising bottoms reflect higher and higher support levels for the security or commodity. If combined with a series of ascending tops, the pattern is bullish.

FIGURE R.7—RISING BOTTOMS

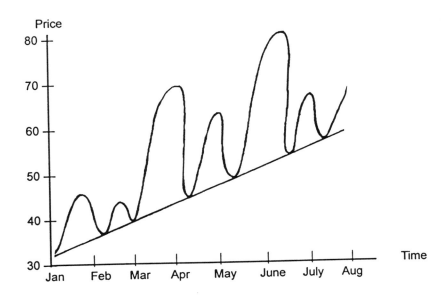

ROOFLINE

A horizontal line depicting the top of a price pattern or formation.

ROUND TRIP

The full cost of a complete transaction—buy and sell—in a stock or commodity.

RYAN INDEX

An unweighted average of total return based on the prices of the most recently auctioned treasury notes and bonds with maturities of 2 to 30 years. (December 31, 1979 = 100). Developed by the Ryan Financial Strategies Group, it is made up of the current 2-, 3-, 4-, 5-, 7-, 10-, and 30-year Treasury issues that are auctioned by the Treasury on a regular schedule. New Treasury notes have maturities that range from 2 to 10 years; Treasury bonds have 30-year maturities.

Price and income return are calculated for each maturity on the basis of market price changes and actual accrued interest. Each of the seven active maturities comprises a subindex; these are averaged to obtain the composite Ryan Index level, whose base level was set at 100 on December 31, 1979.

It is published weekly in *Barron's*.

For yield comparisons, the index should be compared with similar indexes for different securities, such as the *Lipper Convertible Securities Indexes*.

Caution: When comparing indexes, note the difference in the base period. Also watch for trends.

S

SAUCER

A saucer (bowl) pattern (Figure S.1) is a very slow gradual change in trend direction either upward or downward. Volume diminishes with the market gradually turning, and then gradually increases as the new direction starts to hold.

FIGURE S.1—SAUCER

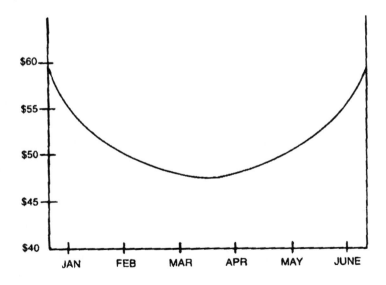

SCHULZ'S RATIO

See Breadth Analysis

SEASONAL MOMENTUM

Progressive seasonal momentum explains how a market moves through its various stages. Each seasonal cycle passes through four stages (Figure S.2).

FIGURE S.2—SEASONAL CYCLE STAGES

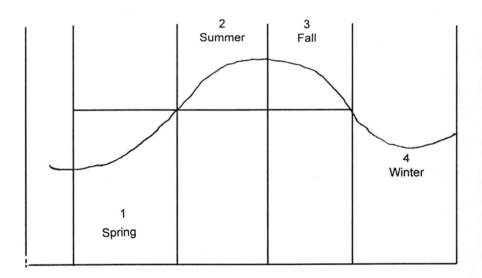

The first stage (spring) takes place after downside momentum has reached its maximum. At this point, the series turns higher but is still lower than its equilibrium level. The second phase (summer) is indicated when the series crosses above the zero reference line. The third stage (fall) begins when it peaks out above zero. The last stage (winter) is triggered when the indicator again falls below the equilibrium level.

From a technical investment perspective, we have the following:

Season	What Happens
Spring	Accumulation
Summer	Markup
Fall	Distribution
Winter	Markdown

A diffusion index may be computed from the position of the seasonal momentum of its components such as industry categorizations for a stock market average and commodity prices for a commodity index. This seasonal momentum method has a number of benefits. It identifies the cycle stage (e.g., accumulation, distribution) and buying or selling opportunities.

The selection of a time span is crucial for momentum indicators. Seasonal momentum may be used for stocks, bonds, and commodities, in both domestic and international markets. For example, a series based on a smoothed 26-week rate of change is much less significant in a long-term investment strategy than a series tied to a 52-week rate of change. This methodology can be applied for monthly, weekly, or daily information. A monthly series is more accurate than a weekly or daily one.

In terms of seasonal momentum for the stock market, a high reading for the spring series signals momentum for stock groups. They go from below zero and rise for a possible major advance.

In most cycles, a chronological sequence exists as most groups go from spring to summer on through fall to winter. A low in the bear market usually takes place about when winter momentum peaks and moves under its 9-month moving average. There is confirmation when the S&P Composite penetrates its moving average.

Spring momentum usually peaks out at the first intermediate-term peak in the bull market. However, it is not a bearish indicator. It means that most groups are going from the accumulation stage in the spring to the markup phase in the summer. It can be a bearish indicator when a substantial number of groups are moving from spring back to winter.

Possible weakness is indicated when summer momentum begins to turn down. However, a sell decision does not arise because the market typically moves sideways or higher after a summer momentum peak. It does imply a more selective environment as the smoothed momentum for more groups goes to the distribution stage in the fall.

The S&P Composite occasionally declines in the transition from summer to fall. However, it is more typical for downside momentum to increase as the cycle goes from fall to winter. That is the point where momentum indicators for most groups cross below their zero reference lines.

A major buying opportunity takes place when winter momentum peaks and begins to turn down. The higher the peak, the more potential for upside movement, since a move from winter momentum flows into spring. A high and falling level in winter momentum shows that a substantial number of groups have the potential to move into the spring position, that is, to move to the point where they are most likely to increase.

The winter momentum series usually goes steadily up toward a peak and then reverses direction. Reversals from high readings typically indicate that the downward movement in the broad market has reversed. On occasion, winter momentum temporarily peaks out but the overall market does not bottom. It is important to look at other technical indicators besides the seasonal momentum measures, and it is crucial to wait for the momentum indicator to go above or below its moving average before determining that there has been a reversal in trend.

A new bull market may be indicated by a downtrend reversal in spring momentum. However, because lead times can be very long, an upward movement in spring momentum is no assurance that there will be a market rally.

A confirming indicator of a major bottom is typically signaled by a trend reversal of summer momentum coupled with or near a peak in winter velocity. The less the summer velocity when there is a reversal, the greater the likelihood of a market increase.

Market tops are much more elusive than bottoms. However, there is typically an advance warning of a reversal in the uptrend of spring momentum. Although lead times may vary, the presumption is that if summer and spring momentum are both increasing, prices will rise. In most instances, the stock market will increase for at least one month after a spring momentum peak.

Market tops usually take place between the summer peak and fall momentum. Even a topping out of fall momentum may not result in a bear market. A decline in fall momentum typically applies to a distribution or topping out phase. In some cases, group momentum swings back to summer, averting a major market decline only if a significant and expanding number of groups fall below their zero reference lines (e.g., move into winter). There is more downside momentum to a bear market.

Care should be taken when there is an Indian summer, as there was from 1984 to 1987. This period is associated with a strong continual linear uptrend. Here, the typical spring-summer-fall-winter sequence does not take place. Rather, summer and fall alternate, enabling the market to regain its internal strength without experiencing a major decline. This is an Indian summer rally.

In consequence, one cannot conclude that a summer momentum peak will always be followed by a market decline. It is crucial to examine the start of the decline in fall momentum to determine whether stock groups flow into winter or back into summer. In making this determination, winter and summer momentum should be tracked. It is bearish if the winter quadrant continues to rise, indicating that more stock groups are dropping in price. This negatively affects the S&P Composite.

SECONDARY OFFERING (DISTRIBUTION)

A secondary offering is a redistribution of a block of securities after time has elapsed since the issuance of securities. The issue price in a secondary offering is typically based on the prevailing market price of the security. A secondary offering occurs privately, not on the exchange.

SELLING CLIMAX

A span of highly unusual volume at the end of a fast declining overall market. It usually involves margin calls. Investors have suffered losses, leading to panic selling with high volume and significantly lower prices. Downward pressure has ended (the damage has already been done) and an upward reversal is beginning. The low at the climax point is not expected to be violated for a long time. A selling climax often occurs at the end of a bear market.

SENSITIVITY

The percentage change in the price of a particular stock relative to the percentage change in the overall stock market.

SENTIMENT INDICATORS

Sentiment indicators evaluate investor feelings toward the stock market, the people behind the prices, and the actions they take. They have their basis in contrary opinion, being founded on the premise that majority opinion is typically incorrect. Sentiment indicators measure bullish or bearish mood.

Emotions affect stock prices. Sentiment analysis indicates exaggeration or underestimation on the part of investors that may be helpful in determining if a trend exhaustion is likely to become exhausted. In other words, the trend and level of sentiment are essential indicators of a market reversal. A worthwhile methodology to predict early reversals is to consolidate both sentiment and momentum into one series.

Because sentiment indicators are vulnerable to institutional changes, they should be considered as a group rather than individually.

Sentiment indicators consider contrary opinion, investment advisory recommendations, mutual fund cash position, investor confidence, short selling, put/call ratio, trading in large blocks, odd lot analysis, and margin debt position.

CONTRARY OPINION INDICATORS

Sentiment indicators should be used to evaluate the consensus view against which a contrary position can be taken. For example, small investors are typically wrong on how the stock market will fare. Thus, money can be made by doing the opposite of what small investors do. Further, the crowd is usually incorrect at key market turning points. If most investors are bullish, stock prices are about to decrease; they will increase if most investors are bearish.

Because the news media reflect majority opinion, they may be used as a contrary indicator. Financial magazines, network financial news, financial books, etc., should all be reviewed.

If too much good news in the media leads to a run-up in stock prices, there is a likelihood the market is too bullish; a drastic decline on bad news suggests the market is too bearish. When a market top or bottom is reached, some sort of reversal in trend is likely.

If a news event that would typically move prices one way or the other does not, such news is already incorporated into the price. Further, if a news story does not have the expected market response, it is most likely in the process of turning.

INVESTMENT ADVISORY SENTIMENT

Investors Intelligence surveys the recommendations in investment advisory reports. They are most bullish near market tops and most pessimistic near market bottoms. To make a profit, take the position opposite that of investment advisory services. For example, the Advisory Sentiment Index developed by A.W. Cohen is a contrary measure. The index is based on a review of approximately 100 stock market newsletters. A determination is made of how many services are bullish or bearish. The premise is that advisory services are usually incorrect about what stock prices will do at major turning points. A 4-week simple moving average of the index will smooth out weekly figures. A bullish reading is less than 37.5% while a bearish reading exceeds 75%. Barron's reports this index weekly.

An Advisory Services Sentiment Indicator that moves below 35% and then above it sends a key buy signal. At market tops a decline from the 80% level to less than 75% level is a warning of impending problems.

MUTUAL FUND POSITION

Mutual fund information useful for tracking activities of the public and institutions is issued each month by the Investment Company Institute (ICI).

Mutual funds must keep part of their portfolio in liquid assets for possible redemptions. Mutual fund cash relative to total assets must be considered. The mutual funds cash/assets ratio equals:

Total mutual fund cash + cash equivalents
Total assets under mutual fund management

This ratio shows how mutual fund managers are thinking. It is both a sentiment and a flow of funds indicator. The ratio is looked at as a contrary measure because past experience shows that professional money managers are usually wrong at market extremes. Thus, a high ratio is bullish, a low one bearish. However, caution should be exercised in making these conclusions because (1) professional managers are sophisticated, and (2) the more cash available in mutual funds the greater the possibility of stock prices increasing when the funds are used.

The mutual funds cash/assets ratio goes in the opposite direction of stock prices because the percentage of cash held by mutual funds increases as market prices of stocks decrease, and vice versa. There are three reasons for this:

1. More cash may be held for public redemptions as stock prices decrease; similarly, in a bull market there is pressure on mutual fund managers to be fully invested and thus have less cash in reserves.

2. As the net asset value of a mutual fund decreases because of declining stock prices, the cash held will increase as a percentage of total assets even without additional cash inflows.

3. As stock prices decline, mutual fund managers become more conservative and thus the percentage of cash held increases.

In general, the cash position of mutual funds relative to net asset value is about 8%.

Arthur Merrill's Funds Net Purchases Index looks at what mutual fund managers are doing to forecast the stock market. Merrill's index considers "small money" purchases and sales of stock as well as total fund assets (stock, bond, and income funds). The information is available from the Investment Company Institute or in *Barron's* (usually one month afterwards). The Funds Net Purchases Index equals:

$$\frac{\text{Purchases of common stock - sales of common stock}}{\text{Total fund assets}}$$

To smooth the result, a 5-month simple moving average (33% exponential) is used. The number is divided by the standard deviation. According to Merrill's study, a positive index greater than 2/3 of the standard deviation from the mean is bullish, a 2/3 negative deviation is bearish.

INVESTOR CONFIDENCE

While a primary bull market is underway, investors feel confident and buy stocks. However, investor enthusiasm typically peaks when the market is reaching its high. This is the time to sell. In a similar vein, investor pessimism usually peaks at market bottoms. This is the time to buy.

Barron's Confidence Index equals:

$$\frac{\text{Average yield on high-grade conservative bonds}}{\text{Average yield on intermediate grade speculative bonds}}$$

The premise of the Confidence Index is that when investors are very positive about the economy, they are more apt to invest in riskier bonds. In consequence, there is an increase in the prices of speculative bonds, resulting in a decrease in their yields. This increases the Confidence Index. The Confidence Index similarly drops when investors are negative, they become conservative, and speculative bond yields rise.

Thus, when investors are pessimistic about the economy they will shift funds from speculative bonds to conservative ones.

There is a perceived correlation between bonds and stocks. For example, when investors are optimistic about economic conditions, besides investing in more speculative bonds they also buy stocks.

Some theorize that a change in the trend in the Confidence Index leads a change in the direction of the stock market by two to four months.

The *Daily Sentiment Index* looks at the consensus of traders as to prospects in the stock and commodities markets. The index was developed by M. Lively, D. Lively, and J. Bernstein of MBH Commodity Advisors. An index above 90% is extremely bullish, indicating an overbought market while an index below 20% is extremely bearish, signaling an oversold market. Thus, a market turn is imminent.

The *Greed Index* developed by Lee H. Idleman is another contrary opinion rule. It measures the greed of investors, measuring greed by bullish sentiment or optimism. The more greedy or optimistic investors are, it says, the greater the probability that the market will fall. The Greed Index comprises ten different factors that are assigned a value from 1 to 10, including portfolio aggressiveness (high technology versus defensive stocks), acceptance of new ideas, ratio of positive to negative comments by investment analysts, and willingness to invest in untested issues. When the index exceeds 60, it is deemed to be bearish. An index below 30 is a buy signal. The index appears in Barron's.

BULLISH MOMENTUM AND CONSENSUS

Buy and sell signals occur when the Bullish/Momentum Index crosses through oversold and overbought zones and then re-crosses on its way back to zero. Almost every sell indicator is followed by a long correction in either a significant sell-off or a long consolidation time period.

On occasion, a possible trend reversal is indicated if the bullish consensus numbers go to an extreme but are unconfirmed by a similar change in momentum. This scenario constitutes an exception to the premise that bulls are attracted by significantly higher prices and bears by significantly decreasing prices. However, if sentiment moves to an extreme but prices do not, the extent of optimism or pessimism is misplaced, and price adjusts accordingly.

SHORT SELLING

The decline in market averages and individual stocks is typically steeper in bear market moves than the advance in bull market moves. Therefore, some believe that profits can be earned faster on the downside of the market than on the upside. However, there are psychological barriers to short selling.

Short covering is buying stock to close out a short sale. In other words, it is the purchase of a stock that the investor previously had sold short. A profit is earned if the stock or commodity is sold at a higher price and bought later (covered) at a lower price. However, in a short sale the investor losses if the stock price increases. To protect himself against unlimited losses, the short seller may set a stop loss order he is willing to absorb on the short side.

The Federal Reserve requires a short seller to have in a margin account cash or securities worth at least 50% of the market value of the stock sold short. Another requirement is that a stock can be sold short only when the stock price has risen. Stocks traded over-the-counter, on the other hand, can be sold short any time. Short sellers pay brokerage commissions on both the sale and repurchase.

Short selling against the box is selling short stock actually owned by the seller but held in safekeeping. This may be done to protect a capital gain in the owned shares and at the same time defer the taxes that would have to be paid if the shares were sold.

Short selling could strengthen the market later because each short seller is by definition a potential buyer. In fact, some investors actually buy a stock they know has a significant short position (interest) because the short sellers will ultimately have to buy back the stock. (Short position means the total short sales outstanding [not yet covered] at a particular stock exchange at a given date.) For example, short sellers may buy back the stock even after a small decline in price. Short sale information is released weekly by the New York Stock Exchange for both members and nonmembers. The figures are published monthly by the *Wall Street Journal.*

A short sale may not be made at any time or price. A short sale can only occur after an up-tick. Most stock exchanges do not allow the short sale of a stock below the last lot traded so as not to feed a declining market.

A short sale is speculative. There is a slight additional tax assessed on short sales. It also involves a margin transaction. Stocks of large companies are usually available for loan at times in your broker's possession. However, it is possible that a stock one wants to sell short may not be available for loan.

Some reasons an investor may sell a stock short are:

- The security is temporarily overvalued.
- There is an expectation of lower earnings.

A *short squeeze* occurs when prices of a security or commodity futures contract begin to rise significantly and many investors with short positions panic by buying back the stocks or commodities to cover their positions and prevent further losses. The sudden surge in buying results in even higher prices, aggravating the situation.

A number of technical indicators gauge short selling activity. These measures include the member short ratio, public short ratio, specialist short ratio, public/specialist short ratio, professional investor's trinity index, short interest ratio, and total short ratio. For all these purposes, total short sales equal member short sales plus public short sales.

Member Short Ratio. The member short ratio is used to gauge market sentiment by looking at how much short selling is being done by New York Stock Exchange (NYSE) members. Members are trading on the exchange floor for themselves as well as for clients. The ratio equals member short sales divided by total short sales.

NYSE members are considered "smart money," so if they are selling short for their own accounts (a high member short ratio), short selling is advised. However, if NYSE members are buying stock (low member short ratio), taking a long position is recommended. A ratio of 82% or more is bearish; a ratio of 68% or less is bullish.

Public Short Ratio (PSR). The public (non-member) short ratio compares how many shares were sold short by the public relative to total short sales. The ratio equals total public short sales divided by total short sales.

This is a contrary indicator meaning that, since the public is considered uninformed, if the public is selling short, we should buy and if the public is buying, we should sell.

A high ratio means that the public is pessimistic, so contrarily we would expect stock prices to increase. The more significant PSR is, the more pronounced the resulting effect. For example, an extremely high ratio indicates exceptional public pessimism, so we can be confident that a rally (correction) will occur. Further, the longer the indicator is bullish or bearish, the greater the likelihood of a market move.

In general, a 10-week moving average of PSR that exceeds 25% is a bullish sign; one below 25% is bearish. A moving average smooths out erratic movements.

Specialist Short Ratio. The specialist will buy or sell for his own account when there is a temporary disparity between demand and supply. Specialists are very knowledgeable traders because they make markets (balancing buy and sell orders) for stocks they specialize in. They attempt to maintain an orderly market. The specialist short ratio equals total specialist short sales divided by total short sales. A moving average of the ratio (e.g., 5 or 10 week period) is often charted. NYSE data are usually used to compute this ratio.

Specialists are usually correct about predicting stock price trends at key turning points. A high ratio (above .60) is bearish because specialists are selling short a lot; a low ratio (below .40) is bullish because they are not.

Public/Specialist Short Ratio. The public/specialist short ratio equals total public (nonmember) short sales divided by total specialist short sales. Because the public is usually incorrect while specialists are correct about stock price direction, a high ratio is bullish, a low one bearish. In general, the ratio works better in pointing to an advance. A 10-week simple moving average may be used to smooth out fluctuating movements. NYSE data is used. Further, the specialists are those on the NYSE.

The best signals occur when there is a reversal in trend in the ratio from an extreme level and then it either violates a down trendline or crosses above the 38% zone.

Professional Investor's Trinity Index. Robert Cross formulated the Professional Investor's Trinity Index. Combining three measures of short positions so as to smooth out fluctuations, it compares the short selling activity of professionals to the public at large. The index is usually computed weekly. The calculation steps are:

1. The three elements of the index are computed as follows:

   Public short ratio $= \dfrac{\text{Public shorts}}{\text{Total shorts}}$

   Member short ratio $= \dfrac{\text{Member shorts}}{\text{Total shorts}}$

   Specialist short ratio $= \dfrac{\text{Specialist shorts}}{\text{Total shorts}}$

2. The three components are combined using exponential averages to achieve smoothing.

According to Ned Davis' research, a high (above 108%) index indicates significant professional shorting relative to public shorting, which is bearish; a low ratio (below 85%) is bullish because the professionals are not shorting much.

Short Interest Ratio (SIR). The SIR, another contrary indicator, equals monthly short interest on the NYSE divided by the average daily trading volume of the previous month.

A high ratio indicates a lot of short selling because of investor negativity about the stock market. This is deemed bullish because the average investor is considered wrong. Additionally, short sellers will later have to cover their shorts, pushing up stock prices. The SIR works best as a bullish indicator after a long-term decline in prices instead of after a long upturn. A low ratio is bearish because investors are optimistic; they too are probably wrong, and less short covering will be needed later.

Studying the SIR, Ned Davis has found that a ratio exceeding 1.65 is bullish but a ratio of less than 1.25 is bearish.

Analysis of the SIR is complicated by arbitrage and option hedging.

Total Short Ratio. The total short ratio equals weekly total short sales divided by weekly total buy/sell orders. A high ratio means investors are pessimistic. Since it is a contrary opinion indicator, when investors are negative about stock prices the opposite will occur, with stock prices increasing.

PUT/CALL RATIO

Option trading activity may assist in forecasting market trends. The put/call ratio is another contrarian indicator. It equals:

Total volume of put options
Total volume of call options

Call volume usually exceeds put volume so the put/call ratio will be below 1.0. The ratio compares the sentiment of bears to bulls. For example, a low ratio means fewer puts than calls. The put-call ratio increases with more put activity due to pessimism around the market bottom.

The ratio is usually based on daily volume data provided by the Chicago Board Options Exchange (CBOE). A very high ratio (above 0.55) is bullish (because the crowd is bearish) while a very low ratio is bearish (because the crowd is bullish).

Using a 10-day simple moving average reduce variability in numbers. In terms of a moving average result, a reading of 0.50 is bullish while that over 0.70 is very bullish.

The call/put ratio is the opposite of the put/call ratio. It equals:

Daily volume of call options
Daily volume of put options

Because option traders are typically incorrect, a high ratio is bearish and a low one bullish.

Open interest is the total number of outstanding options at the end of the day. Put/call open interest analysis was developed by Bob Prechter and Dave Allman. The indicator compares the open interest on puts and calls on the Standard & Poor's 100 Index. A less erratic calculation is to use a 10-day moving average of the ratio. The put/call open interest ratio is more beneficial in signalling intermediate-term than short-term reversals.

If the indicator based on a 10-day moving average is 2 in either direction it signals a key turning point in the market. A market peak is signaled when the ratio goes above the 1.9 percent level and then falls back below it. A market bottom is implied with declines below 0.50.

The put/call option premium ratio (premium is the cost to buy a put or call option) is a near-term sentiment measure. It equals:

Average premium on all listed put options
Average premium on all listed call options

Based on weekly option data, the ratio is a short-term contrary indicator because more investors are incorrect about stock price direction at market turning points. For example, if investors are excessively upbeat about stocks, excessive call prices arise, resulting in higher call premiums which cause a low (high) put/premium ratio—a bearish sign for stocks. Ned Davis considers a bullish reading to be above 95% and a bearish reading below 42%.

In general larger moves occur when put/call volume ratio and the put/call premium ratio agree.

The *Call-Put Dollar Value Flow Line (CPFL)* developed by R. Bruce McCurtain is a long-term sentiment indicator computed in the following steps:

1. Multiply volume by the closing price for all S&P 100 Index (OEX) call options traded on the Chicago Board Options Exchange.

2. Add the products to compute call dollar value.

3. Multiply volume by the closing price for all OEX put options.

4. Sum the products in step 3 to arrive at put dollar value.

5. Determine weekly net call-put dollar value (call dollar value - put dollar value).

6. Compute a running total of the weekly net call-put dollar value.

7. Compute a 43-week single moving average of the cumulative weekly net call-put dollar value flow line (CPFL).

If CPFL closes higher than its 43-week simple moving average, buy stocks and hold them until CPFL crosses below its own 43-week simple moving average. At that time sell short as well as selling existing positions.

The option buy (initial option transaction establishing a long position) call percentage examines open buy call transactions to total call volume. Investor optimism is indicated with a high ratio but a low ratio points to trader caution.

LARGE BLOCK RATIO

A sentiment indicator reflecting trades in large blocks (10,000 shares or more) compared to the total volume traded on a stock exchange (e.g., NYSE). The ratio equals:

<u>Number of large blocks</u>
Total volume traded

The higher the ratio, the more active are institutional investors. A very high ratio may indicate a market top or bottom. This is because institutions (who are smart money) make significant trades at times of extreme overvaluation or undervaluation in equity markets. Thus, the ratio may help in identifying major reversal points.

For a more stable ratio, the large block ratio may be plotted over a 30-day moving average.

ODD LOT ANALYSIS

An odd lot is a buy or sell trade of less than 100 shares. It shows investment decisions by odd-lot traders, who are usually small, inexperienced individual investors. There are several brokers who specialize in odd lot transactions.

As a guideline, an odd-lot transaction for stock at a market price under $40 involves paying an additional eighth of a point above the brokerage commission. If the trade is for a $40 market price or above, the addition is a quarter point.

The cost of odd lot trading is proportionately less on higher priced issues. For example, an extra quarter point is not as significant with a stock trading at $100. Stocks having a low market price (e.g., $15) should preferably be bought in a round lot (100 shares), especially if the security is being bought for short-term purposes.

Odd lot activity is a contrarian indicator. A significant number of odd lot buys is usually deemed bearish while a significant number of odd lot sales is viewed as bullish. The notion is to do the opposite of what small, naive odd-lot investors (who are typically wrong about market direction) are doing because they are not knowledgeable.

The *Odd Lot Balance Index (OLBI)* is the ratio of odd-lot sales to odd-lot buys. It is a market sentiment indicator based on daily NYSE information.

A high OLBI means odd lot traders are selling more than they are buying. In so doing, the odd-lotters are bearish. Smart money should do the opposite. Therefore, a high OLBI reading (over 2.5) is actually very bullish.

By plotting a 15-day simple moving average, the chartist can smooth the daily variability of the index.

A significant amount of odd-lot shorts compared to odd-lot sales indicates a major market low (given odd-lot investor naiveté). Thus, a low ratio of odd-lot shorts to total odd-lot sales points to a market top.

The *Odd Lot Short Ratio (OLSR)*, computed daily, represents another measure of market sentiment. The ratio equals:

$$\frac{\text{Odd-lot short sales}}{\text{Odd-lot buys} + \text{odd-lot sales}/2}$$

OLSR, too, is a contrary indicator. If the ratio is high, it means odd-lotters are pessimistic, so the smart thing to do is buy. If the ratio is low, odd-lotters believe the market will rise, so the smart thing to do is sell. Therefore, a reduction in odd lot short selling activity is bearish.

Odd-lotters are usually reactive instead of proactive. A high odd lot short ratio usually occurs after a significant market decline when one should be buying, not selling. A low odd lot short ratio typically occurs after a long market advance when one should be selling, not buying.

Odd-lot trading data are published in financial newspapers such as the *Wall Street Journal* and *Barron's*. Volume is typically stated in number of shares instead of dollars. Some technicians prefer the SEC *Statistical Bulletin*, in which volume is expressed in dollars.

MARGIN DEBT

Margin applies when an investor in a security pays part of the acquisition price of a security with the balance owed on credit (margin) to the broker. Margin is the difference between the loan balance and the market value of the securities. The securities serve as collateral for the loan balance. As stock prices decrease, so does the collateral value of the underlying securities. Some securities, such as over-the-counter small capitalization stocks, cannot be bought on margin because they are highly speculative. In general, margin-oriented investors are comparatively more knowledgeable than cash investors.

Margin is the least amount of cash that must be put up to purchase a security or a commodity contract. The minimum margin requirement on stocks is 50%. However, the minimum margin to buy a commodity contract is 10%. In a stock, interest is charged on the borrowed funds, while on a commodity account interest is not charged. Margin figures for stocks are published each month by the NYSE.

Example: An investor buys stock worth $100,000. If the margin requirement is 50%, the investor must pay $50,000 in cash. The other $50,000 is borrowed from the broker, using the paid equity as collateral.

Margin debt trends reflect investor confidence (sentiment) and the flow of funds. At the start of a usual stock market cycle, the percentage of margin debt is low. It begins to increase shortly after a market bottom. As stock prices increase, investors become more confident and buy a higher percentage of stocks on margin, giving them leverage to earn a greater return.

In a primary uptrend, margin debt is an important fund source for stocks. However, in a downturn, margin debt is a negative and becomes a source of stock supply.

A primary stock market peak is almost always preceded by higher interest rates, which result in a higher cost to carry margin debt, making it less desirable.

If stock prices are increasing significantly over time, increased speculation entering the market results in a significant increase in margin debt.

Margin debt is a good indicator when looked at relative to a 12-month moving average. A crossover is confirmation of a major trend reversal.

A margin call occurs when the value of a portfolio declines so much that the investor must pay additional money or put up additional securities as collateral to meet the minimum requirements established by the stock exchange or brokerage house. The *exhaust price* is the price at which the broker must sell a client's holding in a security purchased on margin that has declined in price but for which the investor did not have sufficient funds to meet the margin call.

Holders of "troubled" margin accounts having an equity interest below 40% not only have less purchasing power available but are also susceptible to margin calls if stock prices fall further. At first, the margin call process is fairly orderly since most investors have adequate collateral. However, a significant drop in market price may force a lot of investors to sell securities because they cannot meet the margin call. This adds to the supply of stock that must be sold immediately, thus furthering the downward spiral.

A low percentage of debt in troubled margin accounts indicates a strong market.

The trend in margin debt should be compared to the trend in the broad stock market. In general, major changes in the trend of stock prices take place when margin debt bottoms or peaks.

FUTURES MARKET

Sentiment indicators can also be used in the futures market. Futures data are tracked in *Market Vane* (Haddaday Publications, Pasadena, California). Market participants are surveyed weekly by *Market Vane* to determine what percentage of participants are bullish. When a substantial number of participants are optimistic on the futures market, it is likely that they already have long positions, so minimal buying power remains. Thus, prices will probably trend downward. Similarly, when most participants are pessimistic, selling pressure is at its maximum, and hence prices will reverse upward. A moving average may be computed to smooth out weekly variability.

SHAKEOUT

A correction in stock or commodity prices prompting insecure investors out of the market before an upturn commences.

SMOOTHING

A mathematical technique to remove excess variability in the data while keeping a correct evaluation of the underlying trend.

SPECIALIST

A stock exchange member who makes a market in one or more listed issues and execute orders on it.

SPECULATION INDEX

The ratio of AMEX volume divided by NYSE volume. The ratio may be computed from total volume information published in financial newspapers such as the *Wall Street Journal*. Speculation is increasing when trading on AMEX stocks (which are typically more speculative) increases faster than trading on NYSE stocks (typically higher-quality issues). A high rate of speculation is a caution sign because of the potential for greater losses.

SPIKE

A fast, significant up or down price movement in a stock or commodity that resembles an upright spike on a price chart, typically occurring over one or only a few days.

SPREAD

The difference in price of two securities, commodities, or precious metals. A spread relates to purchasing one security (commodity, precious metal) while selling another so as to earn a net gain from the increasing or decreasing difference between the two securities (commodities, precious metals). Spreads are usually computed for options.

Example 1: An investor buys copper and shorts gold, anticipating that copper prices will increase faster than the price of gold.

Example 2: An investor buys a February commodities contract and sells an April contract.

STAG

A speculator who gets in and out of stocks for a quick profit instead of holding for the long-term.

STOCHASTICS

Developed by George C. Lane, stochastics is an overbought/oversold oscillator based on the principle that as prices advance, the closing price has a tendency to be ever nearer to the highs for the period. In a downtrend, closing prices usually appear near the lows for the period. Time periods of 9 and 14 days are usually employed in its construction; stochastics values are then smoothed with a simple moving average. Stochastics indicate overbought and oversold conditions. They can be very useful as a timing aid, particularly when used in conjunction with other technical indicators. Stochastics may be used with industry groups or market indexes.

It is important to watch the indicator as it crosses the 75% and 25% lines.

- *75% line*. When the stochastics indicator goes above the 75% line, the stock is overbought. This occurs if the closing price is near the top of the recent trading range; it signals a possible correction—but does not mean that a top has been formed. When the indicator starts falling, however, the stock would be interpreted as topping out, and a new downtrend (negative breakout) comes when the stochastics indicator crosses the 75% line going downward.

- *25% line*. When the stochastics indicator falls below the 25% line, the stock is oversold. This happens when the closing price is near the low end of the recent trading range; it signals a possible market rally. A bottom forms when the stochastics indicator is below the 25% line, but the stock has not bottomed out until the indicator starts to rise. A new uptrend (positive breakout) is indicated when the indicator crosses the 25% line going upward.

The stochastics indicator may be calculated for short or long periods, but it is more sensitive with shorter time spans. A time span that is too short, however, increases the likelihood of whipsaws from too-frequent signals. A time span that is too long may camouflage important signals by requiring a major price move to generate a trend reversal.

Since many stocks can produce significant moves without breaking the 75/25 line, we have added the ability to plot the difference between the Stochastics and a moving average of the Stochastics. In this case:

- A buy signal is given when the histogram crosses the zero line from negative to positive.

- A sell signal is indicated when the histogram crosses the zero line from positive to negative.

The presence of a second moving average will determine which stochastics curve is displayed.

The formula for stochastics is:

$$\%K = \frac{(C - L)}{(H - L)} \times 100$$

where   %K   = stochastic
              C   = latest closing price
              L   = n-period low price
              H   = n-period high price

In the formula, n usually refers to the number of days, but can also mean months, weeks, or hours; Lane recommends using from 5 to 21 periods.

Figure S.3 shows the Dow Jones Industrials 30 and its stochastic. "Buy" arrows were drawn when the %K line fell below and then rose above the level of 25. Similarly, "sell" arrows were drawn when the %K line rose above, and then fell below 75.

STOCK/BOND YIELD GAP

See Yield Gap.

FIGURE S.3—STOCHASTICS FOR DOW JONES INDUSTRIALS 30

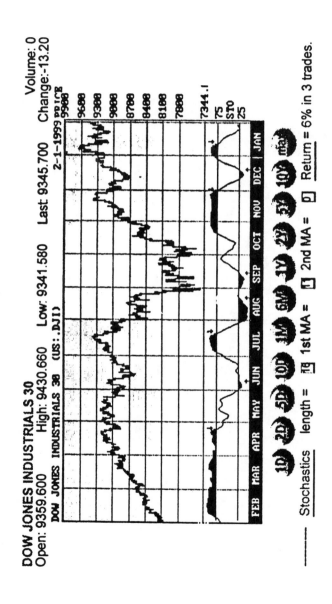

STRADDLE

An option strategy comprised of an equal number of put and call options on the same underlying security, stock index, or commodity future at the same strike price and expiration date. Typically, the combination of options is traded as a group.

STRAP

An option contract comprised of two call options and one put option of the same series with similar exercise prices and expiration dates. The premium paid is less than if the three options were bought separately.

STRONG MARKET

A market characterized by a prepondance of buyers over sellers and an increasing price trend.
See also Weak Market

SUPPORT AND RESISTANCE

Support is the level below the market where buying interest is sufficiently strong to overcome selling pressure. There is support when the consensus is that prices will not go lower. If declining prices meet demand and bounce back, they have reached the support level, which is the price where most investors believe that prices will move upward. Support is thus the actual or potential purchase of stock in enough volume to stop declining stock prices for an appreciable period—or at least temporarily.

Resistance is the area above the market where selling pressure overcomes buying pressure. Resistance occurs when there is a consensus that prices will not move higher. The resistance level is the price where most investors believe prices will move downward. Resistance can be defined as the actual or potential selling of stock in enough volume to stop prices from going higher for an appreciable time or at least temporarily.

Analysis of support and resistance aids is used to estimate the potential extent or duration of a trend. It gives clues as to potential moves in a stock or overall market, and where stocks may run into problems. Penetration of a support or resistance level reveals a change in investor anticipations and a shift in the demand/supply lines.

SUPPORT

The support level is that price level where adequate demand for stock stops a downtrend in price and may even reverse it. The support level is the lowest price in a trading range or the lowest price in a formation whose lower boundary line is horizontal. The support level constitutes a concentration of demand. Support occurs at reaction lows. If market averages go through the support level while certain stocks hold their ground at or higher than their individual support level, those stocks may be more likely to participate in the next recovery.

RESISTANCE

The resistance level is the uppermost price in a trading range, or the highest price in a chart formation in which the top boundary is a horizontal line.

Resistance occurs at reaction highs. The resistance level is the price at which adequate supply of the stock halts an uptrend in price and may even reverse it. In the resistance range, there is a concentration of supply. (The term "concentration" is used because demand and supply are always in balance. However, it is their relative strength or concentration that determines trends.)

A previous top once surpassed becomes a bottom zone in a later downtrend. A previous bottom once surpassed becomes a top zone in a later uptrend.

A breakout above a resistance level coupled with a significant increase in volume makes an upward movement in price more decisive. The longer the sideways movement before a break, the more the stock can increase in price.

In determining how much resistance to anticipate at a particular level, volume, time elapsed, and distance should be taken into account. It may be possible to forecast how far prices will penetrate a resistance range by comparing trading volume on the advance with volume in the original formation of the resistance.

After a series of sharp increases in price of stocks, the investor may reasonably expect a reversal and a significant intermediate decline before resuming an upward move.

Prices can increase easily through a price range where bottoms or congestion areas have not formed in previous downtrends, but if the first resistance level is significantly above, the advance may be exhausted well before it gets to that resistance level. Heavy supply may exist for other reasons at a lower level.

A recovery trend after a bear market panic is typically exhausted long before it goes back up to the last resistance level.

Support and resistance levels can be identified by consistent price reversals at a specific price level. If prices repeatedly retreat from the same level, we have a resistance area. If prices continually advance from the same level, we have a support area.

Analysis of support and resistance is helpful to:

- Identify where a move in stock price is most likely to decelerate or culminate.
- Determine when to buy or sell a stock or switch securities.
- Ascertain at what level a drastic and essential change in demand or supply will occur.
- Determine the magnitude or strength of a move in direction.
- Ascertain how long a trend may occur.

In determining the support and resistance levels, consider:

- The degree of movement before penetration occurs.
- The time expired since penetration.
- Volume.

The greater the volume, the more significant is a support or resistance zone. The more overextended the previous price swing, is, the less support or resistance is needed to halt it. The potency of a support or resistance zone depends on how much time has expired from the formation of the initial congestion to current market developments: A supply that is one year old has greater potency than one set 10 years ago.

One significant deflection off a resistance or support level is much more reliable than a series of low-volume tests. If volume is low, there are not enough market participants to establish a strong resistance or support level.

Over a long period, support and resistance levels tend to gradually adjust, depending on market conditions. For example, a new supply zone may arise after a bear market panic. A panic often downwardly adjusts a previous support level.

A support or resistance level that is tested often and gently over time loses strength, as for example, when prices turn upward on several occasions after a long downtrend, with each upward movement on low volume, followed by a rebound downward. When prices ultimately test the resistance area on significant volume, the selling pressure is not likely to be very strong, having been softened by numerous previous tests.

As a general rule, the longer prices take to turn back through the penetration point, the weaker is the support or resistance. For example, if prices increase after a long downtrend or long sideways movement, resistance will not be as strong as if prices had decreased and picked up again suddenly.

If prices fail to retreat when they go to a resistance (or support) range but stay for a number of days and then push through, there is typically a fast acceleration in price and volume. This implies a decisive break and a sign of a continuous move.

After a breakout from a consolidation or reversal pattern, there may be a short countermove back to the edge of the pattern. If this move is checked at that point, there is support or resistance.

Point and figure charts provide information about support and resistance levels, as Figure S.4 shows for the support level and Figure S.5 for the resistance level. Figure S.6 shows a flag chart pattern to support and resistance levels.

FIGURE S.4—SUPPORT LEVEL EASY TO SEE

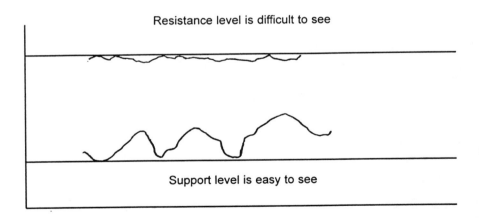

Resistance level is difficult to see

Support level is easy to see

FIGURE S.5—RESISTANCE LEVEL VISIBLE

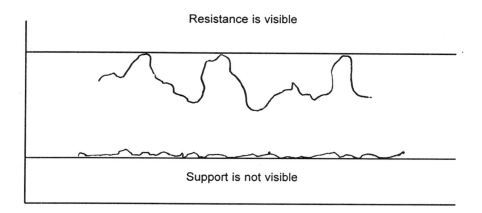

Resistance is visible

Support is not visible

FIGURE S.6—FLAG CHART PATTERN

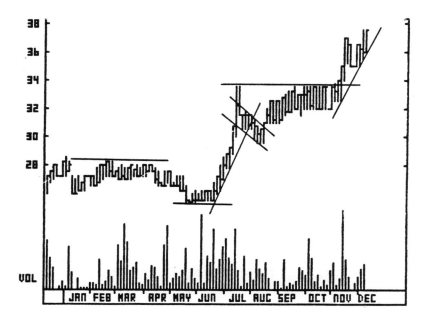

Stocks may go up or down in a zigzag fashion. In an upward move, the reaction after an advance tends to stop at the level of the previous peak. If a stock goes to a new high and then declines on low volume, there may be a buying point at about the peak level reached on the original breakout.

Example: If a stock breaks out from 30 to 40, you may profit by buying on a reaction to 35, assuming that is the support level.

The investor should consider questionable each successive step of an advance. After a stock has made three such moves in the primary direction, an intermediate correction may occur, or a consolidation period. Here are some guidelines for this situation:

- After the first breakout, buy the stock on the reaction to support.
- After the next move, buy the stock on the reaction to the first minor peak, but not on the reaction to the second minor peak.
- Buy on reaction to support.
- Sell on reaction to resistance or a new top.

Support and resistance action in a minor trend is usually clear in daily charts; intermediate and major supports and resistances are clearer in weekly or monthly charts.

If a stock is at the support or resistance level for several weeks, then penetrates that level, closing at a price through that level, you may want to sell the stock.

If a stock makes a minor move in the right direction and returns to the support or resistance level, you may want to sell it if there is a penetration.

In a bull market, if a stock goes to a market top, you may want to wait for an intermediate reaction when the stock drops significantly. Assuming the major trend has not reversed, a buying opportunity may arise at the intermediate support level—typically the top level of the previous advance. The next primary advance may start from this point.

In a bear market, a breakout may consist of a few steps of decline with intervening rallies to minor resistance. An intermediate recovery will follow. Wait for the rally that may consist of a number of minor steps to reach or near a previous intermediate bottom. At this time there may be significant resistance. This is the time to sell the stock short.

In an uptrend, both support and resistance levels rise. Support levels usually hold, while resistance provides temporary halts to upward price movements. Resistance levels are repeatedly broken until there is a reversal in the uptrend.

In a downtrend, both support and resistance levels move lower. Resistance levels usually hold while support levels temporarily halt price movement downward. Support levels are repeatedly broken until there is a reversal of the downtrend.

Figure S.7 shows support and Figure S.8 resistance. Figure S.9 presents declining support and resistance.

FIGURE S.7—SUPPORT

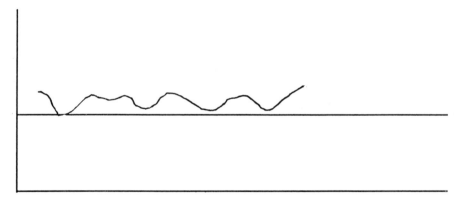

FIGURE S.8—RESISTANCE

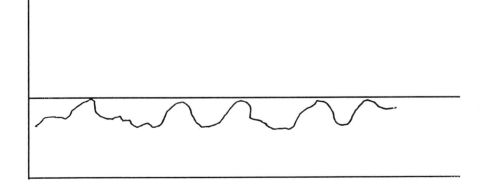

FIGURE S.9—DECLINING SUPPORT AND RESISTANCE

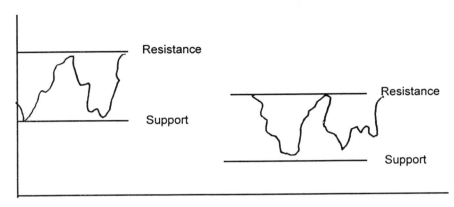

Investor expectations do change as to the support and resistance levels. Other reasons also exist for a change in these levels, such as modified earnings projections, lawsuits, change in management, product line problems, and increased competition. Figure S.10 presents resistance; Figure S.11 depicts a breakout above the resistance level.

FIGURES S.10—RESISTANCE: DOW JONES INDUSTRIAL AVERAGE

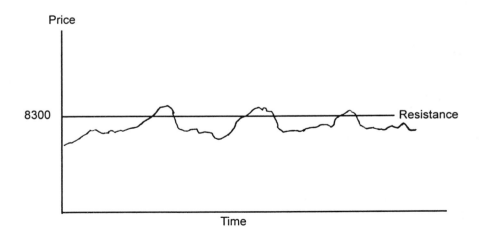

FIGURE S.11—BREAKOUT ABOVE RESISTANCE LEVEL: XYZ COMPANY

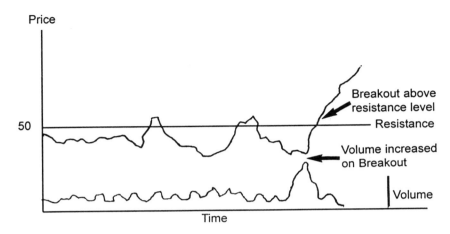

If resistance or support changes roles, the level depends in part on the length of the previous movement.

FIGURE S.12—SUPPORT BROKEN

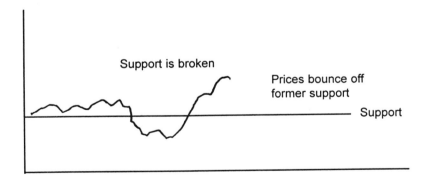

FIGURE S.13—RESISTANCE BROKEN

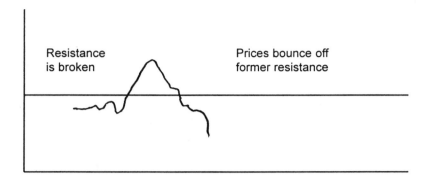

Once support or resistance is broken, there is typically a trend reversal. Figure S.12 shows support being broken. Figure S.13 presents resistance being broken. In an uptrend, for example, resistance levels may turn into support levels after there is a very sharp downturn. This kind of role reversal, in either direction, depends on the following three conditions:

1. The more current the trading at a level, the greater the probability of a role reversal.
2. The longer there is trading at prices near support or resistance, the greater the likelihood of role reversal. An example is when there has been consolidation at either the support or resistance level for a month.
3. The greater the volume at the support or resistance level, the more important that level is, and the greater the likelihood of role reversal.

Figure S.14 depicts rising support and resistance. Figure S.15 shows uptrend role reversal from resistance to support, while Figure S.16 presents downtrend role reversal from support to resistance. Figure S.17 shows support turning into resistance. Figure S.18 presents resistance turning into support.

FIGURE S.14—RISING SUPPORT AND RESISTANCE

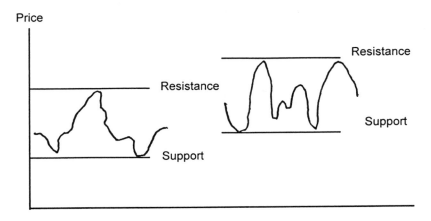

FIGURE S.15—UPTREND ROLE REVERSAL FROM RESISTANCE TO
                SUPPORT

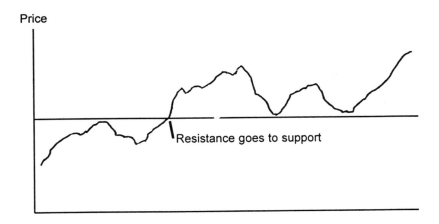

FIGURE S.16—DOWNTREND ROLE REVERSAL FROM SUPPORT TO
RESISTANCE

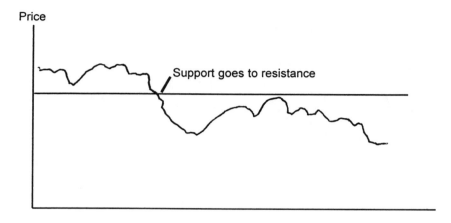

FIGURES S.17—SUPPORT TURNING INTO RESISTANCE

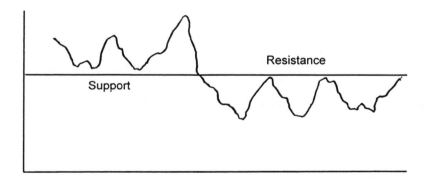

FIGURE S.18—RESISTANCE TURNING INTO SUPPORT

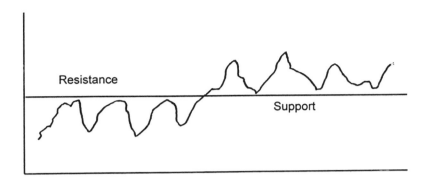

TREND REVERSAL

In an uptrend, the trend reverses when prices are held at a resistance level. A reversal formation such as a double top may then occur and there may be a change in trend direction. In a downtrend, a trend reversal may take place when prices cannot penetrate a support level. In such a situation, a bottom reversal pattern may be formed, and the trend changes direction to the upside.

Note: A trend reversal is not indicated by the initial failure to break through a resistance level in an uptrend or a support level in a downtrend. A fully developed reversal pattern must occur before there is an indication that the trend has changed. Therefore, before deciding to buy or sell, the investor must be assured that a trend is in fact reversing.

Figure S.19 presents a trend reversal at the top, Figure S.20 a trend reversal at the bottom.

TRADER'S REMORSE

When prices return to a support or resistance level after a price breakout it is referred to as traders' remorse. Trader's remorse usually follows the penetration of a support or resistance level as prices retreat to the penetration level. If penetrated, support levels typically provide price resistance, and vice versa. For instance, after the resistance level is broken, investors may question the appropriateness of the new price and decide to sell the stock.

FIGURE S.19—TREND REVERSAL AT TOP

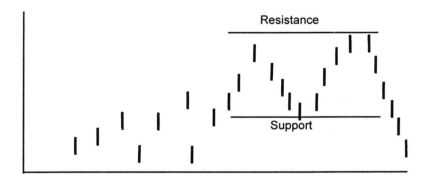

FIGURE S.20—TREND REVERSAL AT BOTTOM

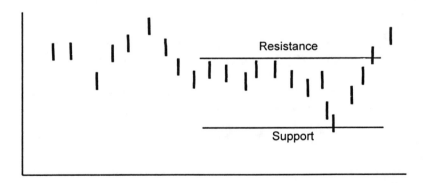

It is important to consider the price action after this remorseful period. There are two alternative possibilities.

The consensus view will be that either the new price is not justified, thereby causing prices to move back to their previous level, or the new price will be accepted by investors so that prices will continue in the direction of the penetration.

If after traders' remorse it is felt that a new higher price is not justified, there will be a false breakout referred to as a *bull trap* (See Figure
S.21). In this case, prices penetrate the resistance level, say $75, resulting
in bulls who anticipate even higher prices. However, prices then decrease
to below the resistance level, resulting in the bulls owning overpriced
stock.

FIGURE S.21—BULL TRAP AND TRADERS' REMORSE

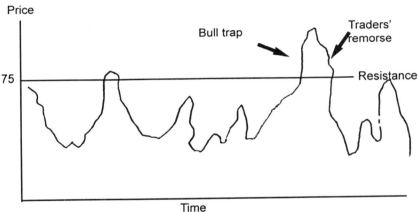

A *bear trap* (see Figure S.22) is also possible. A short-term price
decrease below a support level may prompt bears to sell their stock or sell
short too early. The investor observing the price decline below the support level is misled into believing the price decline will go further.
However, prices then go above the support level, sometimes for an
extended time period, to the detriment of the bears.

The other possibility following traders' remorse is that investors'
expectations change, so the new price is accepted (Figure S.23). If so,
prices will continue to move in the direction of the penetration. In other
words, if the support level was broken, prices will continue downward; if
the resistance level was broken prices will continue upward.

The way to quantify investor expectations after a breakout is to determine how much volume was applicable to the price breakout. If the price
breakout through the support or resistance level is associated with substantial volume and the traders' remorse period with low volume, the
inference is that expectations have changed because investor remorseful-

ness is low. On the other hand, a breakout on moderate volume coupled with traders' remorse period of high volume implies that few investors have changed their expectations so a return to the original prices is justified.

FIGURE S.22—BEAR TRAP AND TRADERS' REMORSE

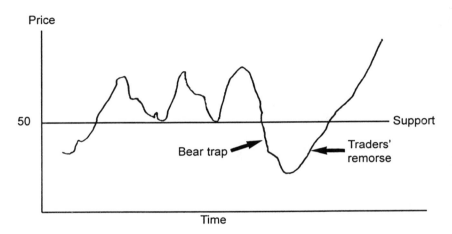

FIGURE S.23—BREAKOUT AND TRADERS' REMORSE

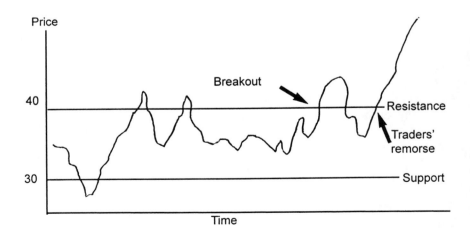

TIRONE LEVELS

Tirone levels are horizontal lines identifying support and resistance levels. They may be drawn based on the *mean method* or the *midpoint 1/3-2/3 method*. The methods identify possible support and resistance levels based on price ranges for a specified time period.

Figure S.24 for ABC Company presents midpoint tirone levels. Average price is depicted by the dotted line. The upper and lower lines divide the range between the highest and lowest prices into thirds.

FIGURE S.24—MIDPOINT TIRONE LEVELS

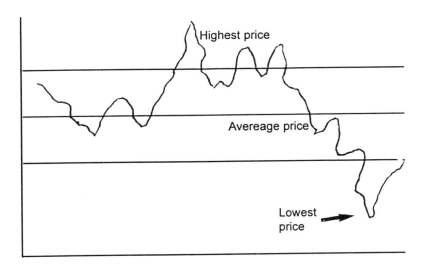

The mean method shows mean levels as five lines, determined as follows:

- *Extreme Low*. The lowest low is subtracted from the highest high. The resulting number is subtracted from the adjusted mean.

- *Regular Low*. Subtract the highest high from the adjusted mean and multiply by two.

- *Adjusted Mean*. The most recent closing price plus the lowest low price plus the highest high price is the adjusted mean. This sum is divided by 3.

- *Regular High.* Subtract the lowest low from the adjusted mean and multiply by 2.
- *Extreme High.* Subtract the lowest low from the highest high and add it to the adjusted mean.

The *midpoint method* involves midpoint levels derived by determining the lowest low and highest high for the period, determined as follows:

- *Bottom Line.* Subtract the lowest low from the highest high. Then divide the result by three. Add the ensuing number to the lowest low.
- *Center Line.* Subtract the lowest low from the highest high. Divide the result by two. Add the result to the lowest low.
- *Top Line.* Subtract the lowest low from the highest high. Divide the result by three. Subtract the ensuing figure from the highest high.

PROPORTION AND RETRACEMENT

In considering support and resistance, proportion should be analyzed. Each action has a reaction. For example, prices of securities are influenced by investor psychology. A study of the levels of support and resistance suggest where a price trend may be halted or reversed. The proportion rule may be applied to both the overall stock market and to individual stocks.

According to the 50% proportion rule, some bear markets have experienced a drop in stock prices by 50%. An example is the bear market covering the period 1937-1938. On the other hand, the proportional principle may show a rising stock market finding resistance after doubling from a low, as happened in the bull market of 1932-1937. Price swings do take place with consistency, so it may be possible to predict reversal points at both troughs and peaks.

In deriving a projection based on proportion, the investor should determine whether the proportion corresponds to a previous support or resistance point. If so, there is a greater likelihood that this zone is a reversal point or at least a temporary barrier.

If a market is moving to an all-time high, trendline analysis may prove useful. The point where the line intersects with the projection based on proportion may indicate the place and time of a key reversal.

Price often moves in proportion. The most typical proportions are 1/3, 1/2, and 2/3. Proportion presumes that each market, commodity, or stock has a character of its own that facilitates prediction as to future movement.

After prices go either up or down for a time, they typically move in the opposite direction, retracing part of the previous move. Afterward, prices continue in the original direction.

Countertrend price moves often move within a percentage range, retracting from one-third to two-thirds of the previous move before resuming the original trend direction.

Some investors consider a retracement of 33% to 50% a chance to buy stocks in an uptrend or sell in a downtrend. The two-thirds guideline is essential. Once prices go past the two-thirds retracement, a trend reversal is highly probable. Percentage retracements are based on price movements in terms of thirds (see Figure S.25).

FIGURE S.25—PERCENTAGE RETRACEMENT

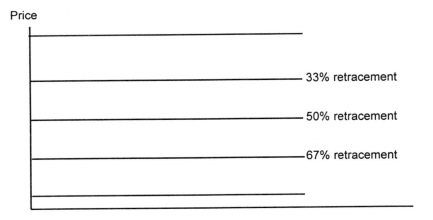

Price

33% retracement

50% retracement

67% retracement

SPEED RESISTANCE LINES

Speed resistance lines are based on the premise that trends are segregated into thirds. A study of how fast an advance or decline takes place may prove enlightening. Speed resistance lines consider both time and price. They measure the rate of change in a trend upward or downward. A rally or drop in prices is measured from the extreme intraday high or low rather than the ending price.

Speed resistance lines serve as important support and resistance areas. The characteristics of speed resistance lines are that:

- A reaction occurring after a rally gets support at the 2/3-speed resistance line. A violation of this line will find support at the 1/3-speed resistance line. If the index goes below the 1/3 line, it is likely that the rising move has been completed and the index will drop to a new low. It may even be below that on which the speed resistance lines were based.
- If the index holds at the 1/3 line, a resistance to an additional price advance may be anticipated at the 2/3 line. If the index goes above the 2/3 line, a new high is probable.
- If the index violates its 1/3 line and subsequently rallies, there will be resistance to that rally at the 1/3 line.

All these in reverse when there is a declining market.

In an uptrend, speed resistance lines are drawn by dividing the vertical distance from the start of the trend to the highest point in the chart. The distance is then segregated into thirds. Two trend lines are constructed (see Figure S.26), the first from the start of the trend to one-third up the vertical distance and the second from the start of the trend to two-thirds the way up the vertical distance. The approach in a downtrend is similar (see Figure S.27).

FIGURE S.26—UPTREND AND SPEED RESISTANCE LINES

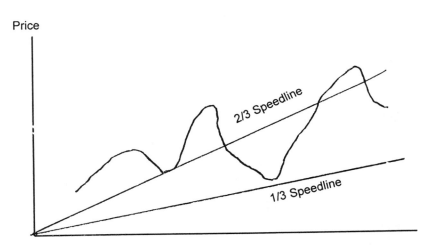

FIGURE S.27—DOWNTREND AND SPEED RESISTANCE LINES

Price

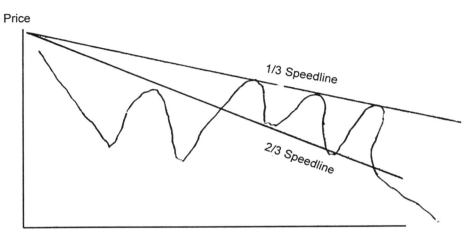

In an uptrend, if prices are at a new peak and are retracing part of the upward move, they will often stop at the two-thirds speed resistance line. If they do not, they will most likely move to the one-third resistance line. If they do not hold there, a trend reversal is indicated.

In a downtrend, a break of the two-thirds resistance line indicates the probability of a rally to the one-third line. If that is penetrated, a trend reversal is likely. In a downward reaction, price may get support when it reaches a line advancing at either two-thirds or one-third of the rate of the advance from a previous trough to a previous peak.

STOP ORDER

With a long position in a stock you do not want it to violate minor bottoms or tops previously made. Hence, you should place a stop order at a certain distance using the minor bottoms (tops) as basing points. In most cases, the most recent minor bottom will be about the same level at the previous minor top so the basing points will usually coincide. Thus, in an increasing trend we consider the most recent minor bottom.

If the stock over three days makes a price range above the entire range of the day marking the bottom, the stop protection may be moved to a new place based on the new basing point.

CONCLUSION

Support is a level in which buying interest is strong enough to overcome selling pressure. A support level is usually identified beforehand by a previous reaction low. Resistance is a price level where selling pressure overcomes buying pressure and a price advance is turned back. Typically, a resistance level is indicated by a previous peak.

T

TECHNICAL ANALYSIS SOFTWARE

Of the numerous personal computer technical analysis software packages, three of the most popular are MetaStock Professional, The Technician, and Telescan Analyzer. We also recommend Compu Trac. Before buying one of these packages for your personal use, be sure to get a demonstration package from the vendor to ensure that it will meet your own needs.

Compu Trac

A comprehensive, powerful, and flexible technical analysis pro-gram for the most serious technician, this software can load 510 periods into memory at a time. This translates into approximately 42 years of monthly data, 9.75 years of weekly data, or 2 years of daily data.

Compu Trac, Inc.
1017 Pleasant St.
New Orleans, LA 70115
1-800-535-7990
(504) 895-1474

Equis International offers three packages:

MetaStock Professional

An excellent and relatively inexpensive charting package with extensive technical analysis capabilities.

The Technician

This fast, easy-to-use package includes more than nine years of historical data and allows you to track 70 popular technical market indicators.

Downloader

An end-of-day data collection program. Via modem, this program automatically collects historical and end-of-day price quotes on securities selected by investors.

EQUIS International
3950 South 700 East, Suite 100
Salt Lake City, UT 84107.
1-800-882-3040
(801) 974-5115

Telescan Analyzer

Allows access to Telescan's online database. Includes charts from one month to 15 years of historical price data and technical information on thousands of stocks, mutual funds, industry groups, and market indexes.

Telescan, Inc.
10550 Richmond Ave., Suite 250
Houston, TX 77042
(281) 588-9700

TechniFilter Plus

A flexible technical analysis program for the most serious technician.

RTR Software, Inc.
19 W. Hargett St., Suite 204
Raleigh, NC 27601
(919) 829-0786

Other technical analysis software packages are described in *The Individual Investor's Microcomputer Resource Guide*, updated and published annually by the American Association of Individual Investors (625 N. Michigan Avenue, Chicago, IL 60611).

DATA VENDORS

Commodity Systems, Inc.
200 West Palmetto Park Road
Boca Raton, FL 33432-3788
(407) 392-8663

Dow Jones News/Retrieval
P.O. Box 300
Princeton, NJ 08543-0300
(609) 452-1511

Reuters Information Services, Inc.
61 Broadway, 31st Floor
New York, NY 10006
(212) 493-7100

Warner Computer Systems, Inc.
1701 Pollitt Drive
Fair Lawn, NJ 07410
(201) 797-4633

TEST

The price of a stock or commodity, or the overall stock market, when it goes to a previous high or low. In effect, this is a test of the validity of a previous trendline.

THIN ISSUE

A stock with relatively little trading volume. As a result, it may experience wide price swings if a significant number of shares are either bought or sold.

THROWBACK

See Pullback or Throwback

TICK

A measure of movement in closing stock prices; a near-term indicator of market strength near the close of the trading day. It is computed during the trading day as the net difference between all New York Stock Exchange (NYSE) stocks with last sales taking place on an uptick (price increase) and those on a downtick (price decrease). For example, a closing tick of +62 means that at their last trade 62 more stocks were rising than were falling.

It is a short-term bullish indicator if tick readings are going from a negative to a positive reading. An example is a change in reading from -100 to +300. TICK readings are of interest to short-term traders. In general, the object is to look for positive or negative trends.

Some consider extreme readings as a contrarian indicator. For instance, an extremely high tick reading may indicate an overvalued stock market and point to a decline to correct it:

Tick Reading	Market Indication
Extremely high	Bearish
Extremely low	Bullish

To smooth out erratic tick movements, a 10-day moving average may be used to closing tick values.

The tick is shown for NYSE stocks throughout the trading day on quote machines. *Barron's* publishes closing tick values for NYSE, AMEX, and DJIA stocks.

TIMING DECISIONS

A good time to buy may be when stock prices are depressed after a prolonged downturn. One theory is to buy on bad news.

A good time to sell might be after a buying climax that is a substantial run-up in stock prices coupled with very high volume. This scenario typically takes place at the culmination of an upward price trend. One theory is to sell on good news.

According to a study done by Arthur Merrill, December is usually the optimum month to invest in the stock market, followed by August and January. The worst months to invest are June and September. He also found that investing at Thanksgiving and selling at the New Year was very profitable. This was followed by investing on July 4th and selling on Labor Day, taking advantage of a summer rally.

Merrill also found that the first six trading days of a month (especially the first two) have a bullish bias.

Stock prices tend to increase on the trading day before a holiday, so this may be the time to buy.

In making a timing decision, the 4% model may be used. The model is based on the percentage change in the weekly close of the Value Line Composite Index (VLCI). It is geometrically based. This is a trend-following measure signaling to stay in the market when there is a major uptrend but to get out of the market or sell short during major market downturns. If the model index increases by 4% or more, buy. If it decreases by 4% or more, sell.

Example: In week 1, the weekly close of the VLCI was 180; in week 2, it was 190. Because the index increased by 5.6% (10/180), the signal is to buy stocks. If in week 3, the weekly close was 182, this would be a sell signal because the percentage decline is 4.2% (8/190). If in week 4, the weekly close was 185, there is no buy or sell signal because the percentage change was only 1.6% (3/182).

TOP

A top occurs when there are successive increases in stock price going to the highest point and then declining for several days.

TOPPING OUT

A reference to a market or a security that is at the end of a period of increasing prices and whose price is now expected to remain fairly constant or decline.

TRADING

1. The buying and selling of securities, usually referring to holding a security briefly to secure short-term profits. An in-and-out trader is one who buys and sells the same security in one day, attempting to gain from a sharp price move.
2. *Outright trading* is having a long or short position in expectation of a price increase or decrease so as to make a profit on the price change.

TRADING SYSTEMS

As soon as a trader completes technical analysis of the market, he should know whether to buy or sell. A trader must decide where and how to enter the market; the decision is founded on a combination of technical guidelines, monetary guidelines, and the kind of trading system to use.

Short-term trading is a popular technique in futures trading and other areas where the timing of entry and exit points must be timed perfectly. The time frame spoken about here usually relates to hours and days rather than months and weeks.

Traders face problems in relation to market breakouts. They can take a position while anticipating a breakout, take a position during the breakout, or wait for the market reaction after the breakout. A trader could use any one of these approaches individually or all three together. A trader who reacts in hopes of an upwards breakout will realize a better price if in fact the anticipated breakout does occur. However, the probability for making a poor trade is increased.

Another trading approach is to wait for the actual market breakout, thus substantially increasing the chances of success. The downside is that the entry price is higher.

The next trading strategy is waiting for the market pullback after the breakout (if it occurs). Unfortunately, this may not happen often in the most volatile markets. This last trading approach is the most conservative. Traders should keep a watchful eye on the trading range of an instrument where prices tend to move in a flat, horizontal pattern.

The prudent trader will use many different trading systems or a combination of systems and approaches to obtain the desired results.

Most successful trading approaches rely on three important elements:

1. *Tactics:* Timing is the most important factor.

2. *Dollar Management:* Includes reward to risk ratios and how much money to invest.

3. *Price Forecasting:* What direction is the market moving in?

All of these things should be factored in when choosing a trading system.

TRADING VOLUME

See Program Trading, Volume

TRADING VOLUME GAUGES

LOW-PRICE ACTIVITY RATIO

The ratio compares high-risk (speculative) stocks to blue chip (low risk) stocks. This measure is published in *Barron's*. It equals:

Volume of low-priced speculative securities
   Volume of high-quality securities

Example: The volume of low-priced speculative stocks is 15 million shares and of high-quality stocks is 140 million shares. The ratio is thus 10.7%.

The market top may be indicated when the volume of low-priced securities increases compared to the volume of blue chip stocks. A ratio of 7.6% indicates a market peak, which is the time to sell securities. A ratio of less than 3% indicates a market bottom, which is the time to buy, since stock prices are expected to increase.

Caution: The definition of a blue-chip or speculative stock is subjective. What is speculative for one person may not be so for another.

See also Volume

NET MEMBER BUY/SELL RATIO

The volume of shares bought compared to the volume of shares sold by members of the stock exchange. These members include specialists in securities and floor traders. The net member buy/sell statistic may be found in Barron's. The ratio equals:

Volume of securities bought by members
   Volume of securities sold by members

Example: If shares bought by members of the stock exchange are 60 million and those sold are 45 million, the ratio is 1.33.

Members of the stock exchange are considered "smart money" because they have significant expertise in securities. If they are net buyers (buyers minus sellers), this is a bullish sign. If they are net sellers, this has a bearish connotation. The trend in member activity over time is a reflection of the direction of market confidence.

Caution: The specialists and floor traders in a stock may be wrong. They are not infallible. For example, they did not expect the stock market crash in October 1987.

ODD-LOT THEORY

An odd lot is a transaction involving fewer than 100 shares of a security. It is usually done by small investors. Odd-lot trading reflects popular opinion. The odd-lot theory rests on the rule of contrary opinion; in other words, an investor determines what others are doing and then does the opposite.

The odd-lot index, a ratio of odd-lot purchases to odd-lot sales, is usually between 0.40 and 1.60. The investor also may look at the ratio of odd-lot short sales to total odd-lot sales, and the ratio of total odd-lot volume (buys and sells) to round-lot volume (units of 100 shares each) on the NYSE. These figures can substantiate conclusions reached in analyzing the ratio of odd-lot selling volume to odd-lot buying volume. Figure T.1 shows a chart of the ratio of odd lot purchases to sales (plotted inversely).

FIGURE T.1—RATIO OF ODD LOT PURCHASES TO SALES (PLOTTED INVERSELY)

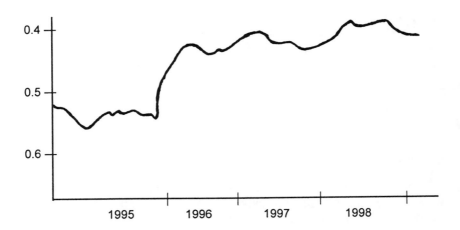

There is also an odd lot short ratio, which equals:

$$\frac{\text{Odd-lot short sales}}{\text{Average of odd-lot purchases and sales}}$$

Example: Odd-lot purchases were 1,500,000 shares while odd-lot sales were 3,000,000 shares. The odd-lot index is therefore .50. The index last period was 1.2. The investor should now buy securities because odd-lot traders, who reflect popular opinion, are selling.

Odd-lot trading data are published in *Barron's*, *Investor's Business Daily*, the *New York Times*, and the *Wall Street Journal*. Volume is typically expressed in number of shares. The Securities and Exchange Commission Statistical Bulletin, another source of data, refers to volume in dollars. Figure T.2 shows odd-lot trading as published in the *Wall Street Journal*.

FIGURE T.2—NYSE ODD-LOT TRADING

Shares in thousands

Daily	Feb 5	8	9	10	11
Purchases	4,522.3	3,540.3	3,407.3	4,309.7	3,818.3
Short Sales	2,169,803	180,329	180.469	164,230	165,792
Other Sales	5,5,282.0	4,278.1	4,382.5	3,942.9	3,597.4
Total Sales	5,451.7	4,458.5	4,563.0	4,107.2	3,763.2
z-zsActual Sales					

Source: Barron's, February 15, 1999

According to the odd-lot theory, the small trader is right most of the time but misses key market turns. For example, odd-lot traders correctly start selling off part of their portfolios in an up market trend but, as the market continues to rise, they try to make a killing by becoming significant net buyers. This precedes a market fall. Similarly, it is assumed that odd-lotters will start selling off heavily just before a bottoming of a bear market. When odd-lot volume rises in an increasing stock market, the market is about to turn around.

The theory: The investor should buy when small traders are selling. Similarly, the investor should sell when small traders are buying.

Caution: Stock market research does not fully support the odd-lot theory.

SPECULATION INDEX

AMEX volume divided by the NYSE volume. The ratio may be computed easily from total volume information available in financial newspapers such as *Barron's* and the *Wall Street Journal.*

Speculation is growing when trading in AMEX stocks (which are typically more speculative) is heavier than trading on NYSE stocks (typically higher-quality issues).

Caution: When there is heavy speculation by traders, there is the potential for greater losses.

UP-TO-DOWN VOLUME RATIO

The number of advancing issues relative to declining ones. It is usually based on a 10-day or 30-day moving average. Up-to-down volume ratio may be found in *Barron's* and the *Wall Street Journal.*

The ratio aids in determining whether accumulation or distribution is occurring. It is helpful in predicting market turning points. For example, a bull market continues only where buying pressures remain strong.

Caution: Even if the up-to-down ratio is increasing, the market still may be headed for future falloff in stock prices due to an overvalued situation.

TREND

A general direction (upward, downward, sideways) in price or volume movement of a stock, bond, commodity, or the overall market. An accelerating trend is one having a slope, up or down, that is increasing sharply. A downtrend is a series of descending peaks and troughs while an uptrend is a series of ascending peaks and troughs. Trendless describes a flat, sideways market. A trending market is when prices go in one direction, typically closing at an extreme for the day.

An uptrend line is one drawn through two or more ascending bottoms; a downtrend line is drawn through descending tops. A violation of an up or down trendline often is a sign of a change in trend.

A primary trend is the predominant movement of a market, whether it be for stocks, bonds, or commodities. A secondary trend is a market movement in security prices in the appropriate direction of the primary trend. A secondary trend is a temporary interruption in a bull or bear market.

A trend may also apply to other financial measures, such as interest rates.

TRENDLINE

A line drawn by technicians to chart the historical direction (movement) of a stock or commodity so as to aid in making future price predictions (see Figure T.8). A trendline connects the lowest or highest prices a stock or commodity has reached within a specified period. A downtrend or uptrend is indicated by the angle of the trendline. A new direction is indicated when price rises above a downward sloping trendline or falls below a rising uptrend line.

FIGURE T.3—TRENDLINE

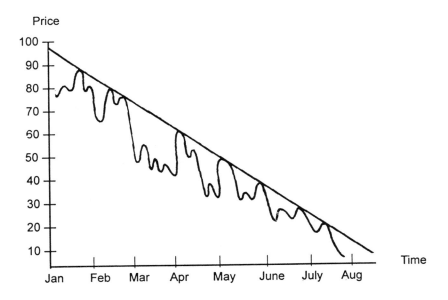

Channel lines are straight lines drawn parallel to the basic trendline. In an uptrend, there is a slant upward to the right above price peaks; in a downtrend the slant is downward below troughs. In most cases, prices meet resistance at rising channel lines and support at declining ones. In other words, a trendline is drawn through a series of ascending bottoms (troughs) or descending tops (peaks). The line moves in the same direction as price.

A buy indicator may be posted when stock price breaks out to the upside of the trendline. The buy signal is more pronounced when it is coupled with a significant increase in volume.

Trendline penetration reliability refers to verifying that a trendline has been broken by considering the shares traded when and after penetration and the degree of penetration.

When the trendline is broken, a trend change may be indicated. For a valid trendline to be broken, it must be penetrated for a number of days. This is referred to as time filter. Greater meaning is associated with a break in the trendline if:

- The trendline is viable longer.
- The trendline has been touched more often.
- The angle of the trendline is less steep

Thus, a trendline is a straight line in a chart between the reaction low in an uptrend and high points in a downtrend. A trendline may be used to identify resistance and support levels.

TRIANGLE

A chart pattern with two base points and a top point formed through the connection of the price movements of the stock with a line. In the usual triangle form, there are right apex points; a reverse triangle has left apex points. With the usual triangle, a series of two or more rallies and price declines occurs. Each peak is below the previous peak. Each bottom is below the previous bottom. With this triangle, there is a sloping pattern that may signal a breakout. When a stock price breaks out of the triangle formation, either higher or lower, the stock price is likely to continue in that direction.

TRIN

See Arms Index.

TYPES OF STOCKS

There are many different types of stock available to the investor. Which you should buy will depend on your particular circumstances and goals. The various types of stocks are described below.

BELLWETHER STOCKS

Securities that reflect market conditions, such as Procter and Gamble.

CYCLICAL STOCKS

Stocks whose price movements follow the normal business cycle. Market prices of cyclical stocks, which include construction and airline securities, increase in expansion and decline in recession. Therefore, these stocks are a bit risky.

INCOME STOCKS

Stocks of companies that pay high dividends and have a somewhat stable stream of earnings. Income stocks are attractive for those who desire less risk and high current income instead of capital growth. A good example would be utility companies. Income stocks give the maximum stable income to satisfy the investor's present living needs.

GROWTH STOCKS

Stocks of companies that exhibit higher rates of growth in operations and earnings than other companies. They typically have higher price/earnings ratios. These stocks normally pay little or no dividend. Growth stocks usually increase in price more quickly than others, but may also fluctuate more, making them more risky. They are popular with investors who seek capital appreciation rather than dividend income.

GLAMOUR STOCKS

Securities with a significant investor following and steadily increasing revenue and profits over the long-term. In a bull market, glamour stocks perform better than the market averages.

DEFENSIVE STOCKS

Often called counter-cyclical stocks, they usually remain stable during periods of contraction. In a bear market, defensive stocks do better than the overall market. They have a somewhat lower return but are safe and consistent. An example would be consumer goods stocks.

PENNY STOCKS

Stocks issued by companies that are financially weak and risky; their market price is usually below $1 per share.

BLUE CHIP STOCKS

Common stocks such as General Electric and Merck that provide an uninterrupted flow of dividends and have strong long-term growth prospects. These stocks have low risk and are less susceptible to cyclical market changes than other stocks. They are of very high quality with a strong profit and price background.

SPECULATIVE STOCKS

Stocks with the potential for large profits but with uncertain earnings potential. These are bought by investors willing to take larger risks for the prospect of high returns.

Speculative stocks have high price fluctuations and price/earnings ratios. An investor investing in speculative stocks may experience significant losses. Mining and biotechnology stocks are typical speculative stocks.

HIGH FLYER STOCKS

High-priced and unusually speculative securities. They have sharp upward and downward movements in stock price in a short period. An example is a new, unproved technological company.

PERFORMANCE STOCKS

High-growth securities that investors believe will substantially increase in market price. Dividend payout is minimal or nonexistent.

WIDOW-AND-ORPHAN STOCKS

Very safe securities that pay high dividends. They are non-cyclical and stable in market price. An example is Bell Atlantic.

WALLFLOWER STOCKS

Securities that are out of favor with investors. They have lower price-earnings ratios.

YO-YO STOCKS

Securities whose price is volatile, with rapid increases and decreases.

U

UPTICK

Uptick is a stock traded at a higher price than the previous price.
See also Downtick.

VALLEY FLOOR

A horizontal line drawn through the bottom point between two consecutive tops. It is V-shaped.

VALUE AVERAGING

Value averaging was formulated by Michael Edleson. The idea is to make the value of an investor's stock holdings increase by a set amount each period (e.g., $10,000) rather than investing a fixed dollar amount, as is the case with dollar-cost averaging. Investors will achieve higher returns at lower per share prices. If the investment increases in value too much, part of the investment should be sold.

VERTICAL LINE CHARTING

A type of charting on which low, high, and closing prices of a security or market are depicted on one vertical line with the ending price shown by a horizontal mark (see Figure V.1). Each vertical line represents a day. The trend is shown over a stated period (e.g., months, years). If the stock or market is repeatedly closing near its low or high of the day, this shows whether market action is weak or strong, and indicates if prices will likely advance or decline.

FIGURE V.1—VERTICAL LINE CHART

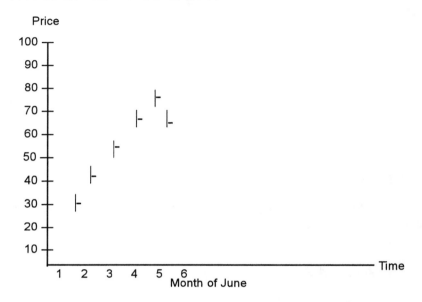

VERTICAL SPREAD

A strategy where an option is bought at one strike price while at the same time another option is sold on the same class at the next higher or lower strike price. The expiration dates of both options are the same.

VOLATILITY

Fast and extreme variability in the price of a security or commodity. Volatility in stock prices may be due to many factors, such as instability in earnings, economic uncertainty, thinly traded security, and erratic economic or political conditions. A measure of stock price variation relative to the overall market is Beta.

VOLUME

The number of shares of stock, bonds, options, or futures contracts traded over a designated period (e.g., daily, weekly, monthly). Advancing volume is the total volume for all stocks increasing in price; declining volume is the total for all stocks decreasing in price. To remove variability elements it may be advisable to smooth this measure with a moving average (e.g., 5 days).

Volume reflects the intensity (strength) of a stock or commodity. Volume also provides an indication of the quality of a price trend and the liquidity of a security or commodity. High volume means greater reliance can be placed on the movement in price than if there was low volume, because heavy volume is the relative consensus of a large number of participants. High volume indicates an active market; in an active market, the spread between bid and asked prices is usually narrower.

High volume is often characteristic of the initial stage in a new trend, such as a breakout in a trading range. Before a market bottom, investor nervousness leads to panic selling, a characteristic of which is high volume. High volume is also attributable to a market top when strong buyer interest exists.

Low volume often exists during an unsettled period, such as at a market bottom. Low volume reflects a lack of confidence that is usually indicative of a consolidation period when prices are within a sideways trading range.

A sizable increase in volume may point to a breakout (start) or climax (culmination) of a move, which may be temporary or final. In a rare case, it may represent a *shakeout.*

Volume typically follows a trend, expanding on rallies and decreasing on reactions. Volume indicators are more reliable if formulated from a smoothed rate of change (ROC) or smoothed trend deviation. Volume is useful in ascertaining how strong a change in expectations really is.

It is important to look at the relationship between volume and price. A price move, up or down, that is on higher volume is more significant. Therefore, an analysis of price and volume allows the investor to better interpret the trends in price and any changes thereto. In other words, volume gives an indication of the strength (momentum) of a move in price.

Current trading volume and average trading volume should be compared. (Average daily trading volume for many stocks is published monthly by the *Wall Street Journal*). Average trading volume typically decreases when a stock is in a downtrend, because investors view negatively a stock declining in price. An increasing price is typically coupled with increased volume, but the price can decrease without an increase in volume if investors lose interest in the issue. On the other hand, a declining stock price may be coupled with higher volume when, for example, negative news comes out about the company.

The significance of a change in volume is related to the associated price trend or pattern. For example, a good time to buy stock is when there are simultaneous price and volume increases. Table V.1 provides general rules for volume analysis.

TABLE V.1—GENERAL RULES IN VOLUME ANALYSIS

Volume	Price	Interpretation
Increasing	Increasing	Bullish
Decreasing	Decreasing	Bullish
Increasing	Decreasing	Bearish
Decreasing	Increasing	Bearish

Volume should be evaluated in appraising market strength or weakness. If volume is increasing, whether prices are going up or down, it is probable that prices will continue their current trend. However, if volume is decreasing, the current trend will probably not continue and a reversal may be imminent.

A strong uptrend usually has more volume on the upward legs; similarly, a strong downtrend will have more volume on the downward legs. After the trend ends the corrective leg usually has lower volume. A downtrend may nevertheless be extended whether average trading volume increases, decreases, or is static.

Volume is relative in that it usually is greater approaching the top of a bull market than near the bottom of a bear market. Further, trading volume typically increases and continues higher than average in an uptrend, but is below average during a downtrend.

Trading volume typically goes up as the price breaks out to the upside of a pattern or formation. In this case, a significant increase in volume is a strong buy signal. However, volume is an indicator of a trend reversal if it goes in a direction contrary to a prevailing trend.

Volume/price analysis, on balance volume, upside/downside volume ratio and line, volume up days/volume down days, cumulative volume index, trade volume index, positive volume index, and volume reversal are among the ways volume can be analyzed.

See also Program Trading, Trading Volume Gauges

VOLUME/PRICE ANALYSIS

Volume is a useful confirmation of price movement. It often leads price, giving an advance indicator of a possible price trend reversal. A rally reaching a new price high on increasing volume but with an overall activity level lower than a previous rally is suspect. It warns of a possible trend reversal (see Figure V.2). Further, a rally occurring on contracting volume is suspicious, again warning of a possible trend reversal.

FIGURE V.2—INDICATION OF A POSSIBLE TREND REVERSAL

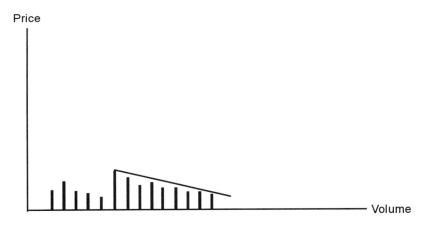

Sometimes price and volume expand slowly at the same time, then go into an exponential rise with a final blow-off stage. After this, both volume and price fall off significantly. This *exhaustion move* has attributes of a trend reversal. The importance of the reversal depends on the degree of the previous advance and the extent of the expansion in volume.

If after a long decline prices react to a level just above or below the previous trough and the volume on the second trough is substantially lower than the volume on the first, it is a bullish indicator.

A downside breakout from a price pattern coupled with significant volume is a bearish indicator that may confirm a trend reversal.

A dramatic market action at a market top is referred to as a *blowoff*; at a bottom it is a *selling climax*. A blowoff typically takes place after prices have edged higher over an extended period. When the up move ends, prices go up significantly, with a substantial increase in volume.

However, this causes profit taking, resulting in a sudden downturn in prices.

A selling climax takes place at a market bottom after prices have been decreasing over a long period. A final wave of selling causes prices to go significantly lower on substantially increased volume. Then, bargain hunters start to buy, reversing the trend, and sending prices higher. The price increase after a selling climax is typically coupled with declining volume. This is the only instance when it is normal to have decreasing volume with increasing prices. It is expected that the low that is set at the time of the selling climax is not likely to be violated for a long time. The end of a bear market is often associated with a selling climax (see Figure V.3).

FIGURE V.3—SELLING CLIMAX

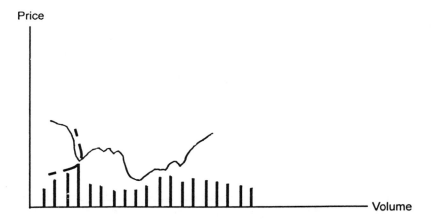

If the market has been rising for many months, an anemic increase in prices but with high volume is indicative of churning. This is a bearish sign. After a decline, substantial volume with minor price change points to accumulation, which is typically bullish (Figure V.4).

A downside price breakout of a pattern or formation is a sell indicator whether or not trading volume increases. If a stock is in an uptrend, average daily trading volume typically increases, since a rising stock price attracts more investors to buy.

FIGURE V.4—SUBSTANTIAL VOLUME WITH MINOR PRICE ACTION

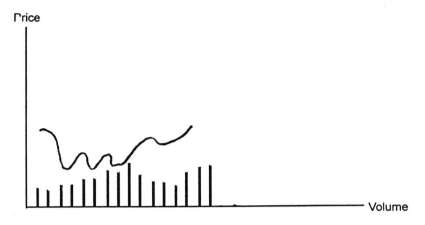

Figure V.5 shows a downtrend moving to an uptrend in price. Figures V.6 and V.7 depict head-and-shoulders bottoms changing to an uptrend. Figure V.8 presents a resistance level broken by an uptrend.

FIGURE V.5—DOWNTREND MOVING TO UPTREND

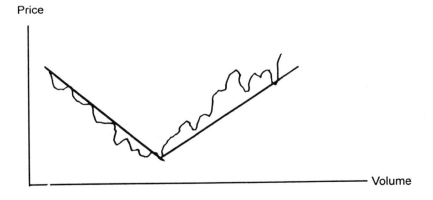

FIGURE V.6—HEAD AND SHOULDERS BOTTOM CHANGING TO UPTREND

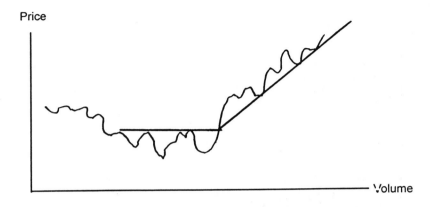

FIGURE V.7—HEAD AND SHOULDERS BOTTOM CHANGING TO UPTREND

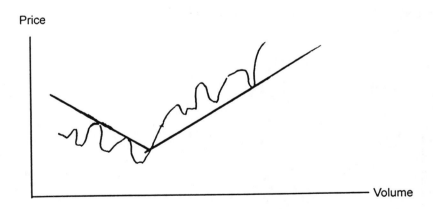

FIGURE V.8—RESISTANCE LEVEL BROKEN BY UPTREND

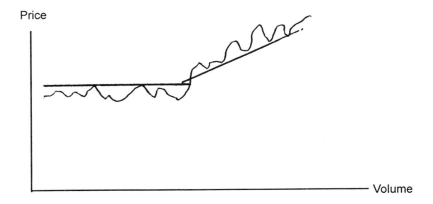

Do not buy a stock that is just beginning an uptrend unless volume is also increasing. If volume has gone up for several days or is unusually high for one or two days, there is a greater likelihood that the uptrend can be continued.

Retain the stock irrespective of variability in volume as long as price continues higher than the uptrend line. If the price falls below the uptrend line by 5% or more, sell the security whether or not there is a change in volume at the time of breakout.

An increase in volume confirms the importance of an upside breakout from a price formation or pattern. The potential of the price increase at breakout is greater the larger the volume increase.

The optimal time to purchase a stock is right after a high-volume breakout to the upside of a bottom formation. This type of breakout is typically followed by an intermediate or long-term price uptrend. The mix of a significant increase in volume with an upside price breakout signals a strong buy. After the uptrend line is established, the stock should be held until the price uptrend is penetrated by 5%.

There may be a price-volume relationship within an elongated trading range. Once a stock falls to a low price, it may be there for an extended period (e.g., years) within an elongated trading range. In this case, trading volume is typically very low, since investors are not interested in stocks that are not doing well. But if the stock price suddenly breaks out above the resistance level with a large increase in trading volume, it constitutes a great buying opportunity.

Figure V.9 shows an elongated trading range where the resistance level is not broken. Figure V.10 presents an elongated trading range in which there is an upside price breakout.

FIGURE V.9—ELONGATED TRADING RANGE

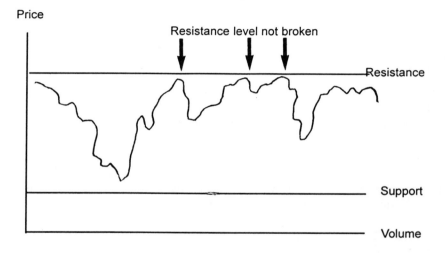

FIGURE V.10—ELONGATED TRADING RANGE MOVING TO AN UPTREND

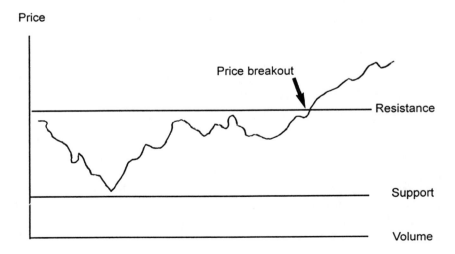

Figures V.11 and V.12 present double and triple bottoms changing to uptrends.

FIGURE V.11—DOUBLE BOTTOM CHANGING TO UPTREND

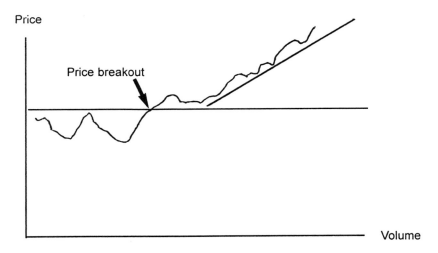

FIGURE V.12—TRIPLE BOTTOM CHANGING TO UPTREND

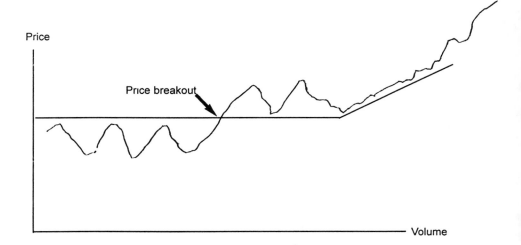

Buy a stock if there is a breakout to the upside along with significant trading volume, and sell it if there is a downside breakout. Once the downside is established, no one can tell how much prices will decline from the associated trading volume.

Figures V.13 and V.14 depict an uptrend becoming a downtrend.

FIGURE V.13—UPTREND BECOMING DOWNTREND

Price

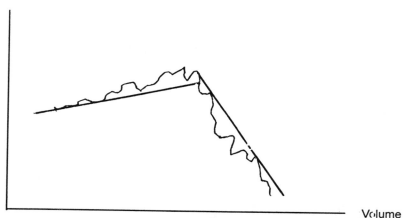

Volume

FIGURE V.14—UPTREND GOING TO DOWNTREND

Price

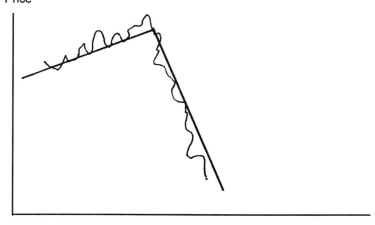

Volume

When stock price forms a triangle, volume typically declines. This reflects a time of investor "wait and see" due to uncertainty. Figures V.15 and V.16 present a symmetrical triangle and a descending triangle changing to a downtrend. Figure V.17 presents a rounding top changing to a downtrend.

FIGURE V.15—SYMMETRICAL TRIANGLE CHANGING TO DOWNTREND

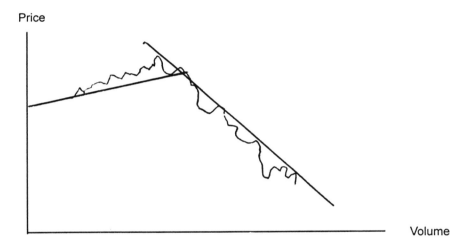

FIGURE V.16—DESCENDING TRIANGLE CHANGING TO DOWNTREND

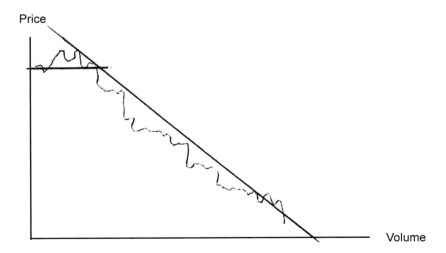

FIGURE V.17—ROUNDING TOP CHANGING TO DOWNTREND

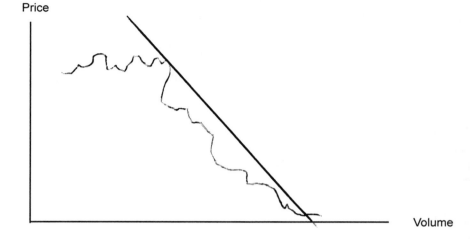

A downside breakout is a sell indicator regardless of the associated trading volume. Downside breakouts arising from head-and-shoulders formations, double-top formations, triple-tops, rectangles, and triangles are all red flags. At a breakout, there is typically an intermediate or long-term price decline.

FIGURE V.18—HEAD-AND-SHOULDERS TOP CHANGING TO DOWNTREND

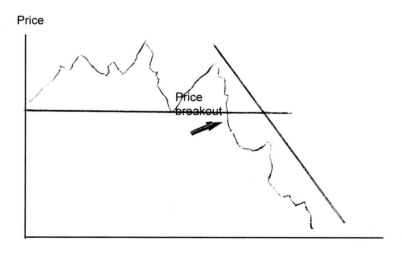

Figure V.18 presents a head-and-shoulders top changing to a downtrend. Figures V.19 and V.20 show a double and triple top, changing to a downtrend. Figure V.21 shows a rectangle changing to a downtrend.

FIGURE V.19—DOUBLE TOP CHANGING TO DOWNTREND

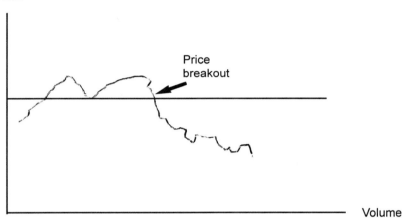

Price

Price breakout

Volume

FIGURE V.20—TRIPLE TOP CHANGING TO DOWNTREND

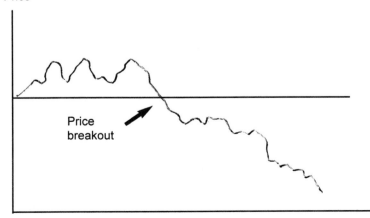

Price

Price breakout

Volume

FIGURE V.21—RECTANGLE CHANGING TO DOWNTREND

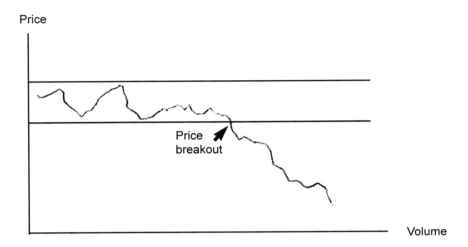

Price

Price
breakout

Volume

ON BALANCE VOLUME

On balance volume (OBV), a model formulated by Joseph Granville, attempts to identify when a security or commodity is being accumulated by many buyers or distributed by many sellers. OBV may be used to evaluate the overall market or a specific stock, bond, or commodity.

OBV is a measure of momentum and a running total of volume. There is a presumption that changes in OBV come before changes in price. When the OBV of a stock is increasing, it assumes that intelligent and sophisticated investors are buying it. Then, as the average investor starts buying the stock, the stock and the OBV will increase significantly.

When the stock price closes higher than the price on the previous trading day, we have up-volume for the current day. If the stock price closes below the close of the previous day, we have down-volume.

A change in stock price that comes before movement in OBV is a nonconfirmation signal. Nonconfirmation can take place at the top of a bull market or at the bottom of a bear market. In a market top scenario, for instance, the stock price increases before or without the OBV.

The OBV line is superimposed on the stock price line of a diagram. It is significant when the two lines intersect. A chart will point to a buy indicator when there is accumulation and a sell is posted when there is distribution.

There is an increasing trend in OBV if the new peak is higher than the previous one or if the new trough is higher than the previous one; in a decreasing OBV trend, the new peak or trough is lower than the previous one. Figures V.22 and V.23 show an uptrend and a downtrend.

FIGURE V.22—UPTREND

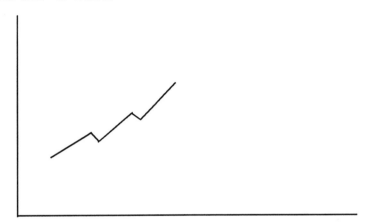

FIGURE V.23—DOWNTREND

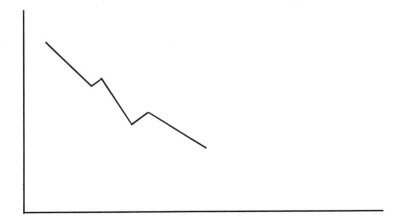

A *doubtful trend* occurs when the OBV is going sideways without successive lows and highs. (See Figure V.24)

FIGURE V.24—DOUBTFUL TREND

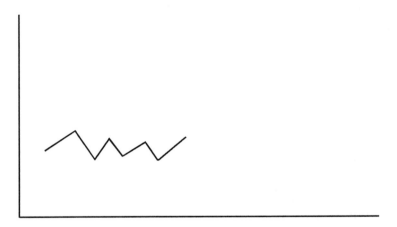

An OBV trend can be broken if the trend goes from an increasing one to a decreasing one, or vice versa. The OBV trend is also broken if the trend goes to a doubtful one for four or more days.

As an indication of buying or selling pressure, the actual amount of on-balance volume is not as important as its direction compared to market price.

There is a breakout when OBV changes to a decreasing or increasing trend. A buy decision should be made with an upside break in OBV, and a sell or sell short decision with a downside break, because OBV usually breaks out before price. If OBV is moving sideways, no decision is necessary. Short-term trading cycle decisions based on OBV can be very profitable.

Example: A stock moves from an increasing trend to a doubtful one for two days before moving back to an increasing trend. The OBV is deemed to have continued in an increasing trend because a doubtful trend must last at least four days.

OBV is computed by adding the day's volume to a cumulative total if stock price closes higher and deducting it if price moves lower. The calculation based on the following scenarios follow:

1. Today's close = Yesterday's close
   Today's OBV = Yesterday's OBV

2. Today's close exceeds yesterday's close
   Today's OBV = Yesterday's OBV + Today's volume

3. Today's close is below yesterday's close
   Today's OBV = Yesterday's OBV - Today's volume

Table V.3 presents a divergence analysis of on-balance volume.

TABLE V.3—DIVERGENCE ANALYSIS OF ON-BALANCE VOLUME

On-Balance Volume	Price	Appraisal
Increasing	Increasing	Bullish
Increasing	Decreasing	Bullish
Decreasing	Increasing	Bearish
Decreasing	Decreasing	Bearish

In studying OBV, if stock price increases to a new high and the OBV does not, it is viewed as a 9negative sign. If price decreases to a new low and OBV does not, the sign is positive.

UPSIDE-DOWNSIDE VOLUME RATIO

The upside-downside volume ratio equals:

Volume of advancing issues
Volume of declining issues

Daily NYSE information is usually used in the computation.

A high ratio is bullish as it reflects buying pressure; a low ratio reflecting selling pressure is bearish. A 10-day moving average should be used to smooth the readings of this ratio. An upside-downside ratio that exceeds 2.3 is bullish and a reading below 0.75 is bearish.

UPSIDE/DOWNSIDE VOLUME LINE

Upside/downside volume data are published daily in the *Wall Street Journal* and weekly in *Barron's*.

The upside/downside volume line is based on the difference between advancing volume versus declining volume of stocks. The line usually rises during market advances and falls during declines. If the line does not confirm a new high or low in the price index, it is a sign warning of a potential reversal in trend.

If the market is advancing irregularly, with successively higher rallies interrupted by rising troughs, the likelihood is that the upside/downside line will be similar. This scenario shows that the volume of advancing issues is expanding on rallies and contracting on declines. If the normal trend of the price/volume relationship is broken, either upside volume is not expanding sufficiently, or downside volume is expanding excessively. Either is a bearish indicator.

The upside/downside line is especially useful when prices are going to new highs and volume is increasing. In this situation, if volume of declining stocks is increasing relative to advancing ones, it means either a slower advance in the upside/downside line or a decline.

If successive highs in the market are coupled with new highs in the volume indicator, it is unlikely that there will be a major sell-off, since market peaks are typically preceded at least one divergence in the upside volume indicator.

A market bottom is indicated when new lows in price are not confirmed by new lows in upside volume. This usually means less downside volume.

Information about the most active stocks is published both weekly and daily. It is beneficial to track the active stocks because they indicate what institutions are doing; they also account for approximately 20% of the volume on the NYSE. Because there is a record of the net price changes of these issues, one can formulate an indicator to confirm movements in the upside/downside line:

If advancing issues over a specified period (e.g., monthly) outnumber declining issues, the oscillator will have a reading above zero.

An index of the cumulative difference between the number of advancing versus declining issues per day should be compared to the advance/decline line. They should move in harmony. If they diverge, a reversal in the overall market trend may take place.

VOLUME UP DAYS/VOLUME DOWN DAYS

Arthur Merrill formulated a measure of volume up days to volume down days derived as follows:

$$\frac{\text{Total volume for the last 5 trading days the stock market went up}}{\text{Total volume for the last 5 trading days the stock market went down}}$$

An index exceeding 1.05 is deemed bullish; one less than .95 is bearish.

Example:

Day	Volume of Shares (in millions)	Stock Market
1	290	Down
2	250	Down
3	305	Up
4	280	Up
5	190	Up
6	260	Down
7	180	Down
8	220	Up
9	310	Up
10	365	Down
11	200	Up

Volume up days/volume down days equals:

$$\frac{200 + 310 + 220 + 190 + 280}{365 + 180 + 260 + 250 + 290} =$$

$$\frac{1200}{1345} = 0.89$$

The reading, 0.89, is bearish.

Advancing volume is bullish because it reflects that more stocks are increasing in price for the trading day, which shows buying pressure. Declining volume is bearish because selling pressure is causing more stocks to decline in price.

CUMULATIVE VOLUME INDEX

A measure similar to OBV is the cumulative volume index (CVI). CVI measures the momentum of the market by determining how much money is going into or out of the market. The index equals the cumulative (running) total plus the excess of advancing volume over declining volume:

CVI = Yesterday's CVI + (Advancing volume - declining volume)

CVI is simply a running total of net upside minus downside volume. Daily NYSE data are typically used in the computation.

Example: The running total of CVI at the start of a trading day was 82. During the trading day advancing volume was 300 and declining volume was 250. The CVI at the close of the trading day is 132 (82 + 50).

The trend, pattern, and slope of CVI are important. As is the case with most cumulative indicators, the trend in CVI is more significant than its actual value.

The divergence between CVI and a market index should be determined. If the market index is at an all-time high but CVI is not, the market index is likely to go down to be consistent with CVI. When the market index is increasing while CVI is decreasing, the indicator is bearish because it implies market weakness; the opposite effect is bullish.

TRADE VOLUME INDEX

The trade volume index (TVI) indicates whether a stock is being bought (accumulated) or sold (distributed). In calculating this index, intraday tick prices are used. TVI presumes that trades at higher asking prices represent accumulation, while trades at lower bid prices represent distribution. Thus, TVI aids in determining if a stock is being bought or sold. A TVI that is trending up indicates that trades are occurring at asking price and buyers are accumulating; a downtrending TVI signals distribution.

There is similarity between TVI and on balance volume (OBV). OBV is good to use with daily prices but it is not suitable with intraday tick prices. Tick prices, particularly for stocks, show trades at the bid or ask prices for long time periods without changing. This results in a flat support or resistance level. At times of unchanging prices, TVI helps identify this volume on either the buy or sell based on the last price change. If prices result in a flat resistance level and the TVI is increasing, prices may breakout to the upside; a flat support level for prices and a decreasing TVI signal a downside breakout.

To compute TVI, add each trade's volume to a cumulative total if prices are increasing and subtract if they are decreasing by a minimum tick value (MTV) amount.

In computing TVI, a determination must be made whether prices are being accumulated or distributed. The following guidelines are used:

- Change equals price minus last price.
- If change exceeds MTV, accumulate.
- If change is less than MTV, distribute
- If change equals or is less than MTV or if it equals or exceeds MTV, the last direction applies.
- When the direction is to accumulate, TVI = TVI + today's volume.
- When the direction is to distribute, TVI = TVI - today's volume.

POSITIVE VOLUME AND NEGATIVE VOLUME INDEXES

The *positive volume index (PVI)* looks at the increase in volume from one day to the next. An increase in volume is attributed to greater interest in stocks by uninformed lay investors. On the other hand, when volume has decreased from the previous day, it is assumed that intelligent sophisticated investors are quietly taking positions. Hence, the index reflects what uninformed investors are doing. However, PVI is not a contrary opinion measure because it does move in the same direction as prices.

The PVI is a cumulative indicator based on a market index or the net advances (advancing less declining issues on the NYSE). The computation may use daily or weekly data. Since prices typically increase with increasing volume, there is an upward trend in PVI.

If PVI is greater than a 52-week moving average, there is a more than 75% likelihood that a bull market exists; if PVI is lower, there is a more than 60% probability that a bear market is taking place.

PVI is computed as follows:

1. If today's volume exceeds yesterday's volume:
   PVI = Yesterday's PVI + <(today's close - yesterday's close/yesterday's close) multiplied by yesterday's PVI>

2. If today's volume is less than or equal to yesterday's volume:
   PVI = Yesterday's PVI

Unlike PVI, the *negative volume index (NVI)* looks at the decrease in volume from one day to the next. It is presumed that a decrease in volume is attributed to "smart money. NVI reflects what intelligent investors are doing.

There is an extremely high probability of a bull market and profitable opportunities when the NVI exceeds its 52-week moving average. However, when the NVI is less than its 52-week moving average, no reliable conclusion can be rendered.

Since decreasing prices are usually associated with decreased volume, there is typically a downward trending NVI.

NVI is computed as follows:

1. If today's volume is below yesterday's volume:
   NVI = Yesterday's NVI + <(today's close - yesterday's close/

   yesterday's close) multiplied by yesterday's NVI>

2. If today's volume equals or exceeds yesterday's volume:
   NVI = Yesterday's NVI

VOLUME REVERSAL

Volume reversal, refined by Mark Leibovitz, theorizes that volume comes before price. Thus, by studying changes in volume—up or down—one can predict changes in stock, commodity, or index futures prices.

A volume reversal takes place when a change from a rally day to a reaction day, or vice versa, is coupled with an increase in volume. (A rally day is one in which the intraday high exceeds the previous day's high or the intraday low is higher than the previous day's low. A reaction day is one in which the intraday low is lower than the previous day's low or the high does not make it up to the previous day's high.) When volume increases and the conditions for a reaction day are satisfied, it is construed as a negative volume reversal and a sell indicator; when volume increases with conditions for a rally day, the positive volume reversal says buy.

The volume reversal approach does not consider inside or outside days. An inside day is one in which the intraday high is the same or lower than the previous day's high or the intraday low is the same or higher than the previous day's lot. An outside day is when the intraday high is higher than the previous day's high or the low is lower than the previous day's low.

The volume reversal approach assumes that the range between the high and low price for a trading day is a superior measure of price. The closing price for the day is not used.

Example 1: The following information is a record of the Dow Jones Industrial Index:

Day	High	Low	Close	DJIA Volume
April 1	9055	8902	9007	107,821
April 4	9067	8931	9004	105,532
April 5	9068	8918	8949	87,353
April 6	8989	8873	8941	120,771

April 4 was a rally day (higher high and higher low) followed by an outside day (higher high and lower low) on April 5. On April 6, there was a reaction day (lower high and lower low) coupled with increased volume. This indicates a negative volume reversal, posting a sell signal.

Example 2: The following record is for the Dow Jones Industrial Average:

Day	High	Low	Close	DJIA Volume
August 5	2292	2260	2268	17850
August 6	2285	2230	2270	18084
August 7	2334	2274	2330	20010

This example reflects a positive volume reversal, indicating a buying opportunity. August 6 was a reaction day (lower high and lower low). The next day, August 7, was a rally day (higher high and higher low), with increased volume.

SUMMARY

The following guidelines apply to the study of volume:

- The market is bullish if a new high occurs with heavy volume. A new high on light volume is deemed temporary.

- A new low price with high volume is a bearish indicator. A new low on light volume is less significant.

- A rally to a new price high on expanding volume but with less activity than the previous rally is questionable. It may point to a coming reversal in trend.

- A rally on contracting volume is questionable. It warns of a possible price reversal.

- If prices advance after a long decline and then go to a level at or above the previous trough, the indicator is bullish when volume on the secondary trough is less than the first.
- If the market has been increasing for a while, an anemic price increase coupled with high volume is a bearish sign. After a decline, substantial volume with minor price changes points to accumulation, typically a bullish indicator.

If program trading accounts for a high percentage of share volume, institutional investors are very active, which may have a pronounced effect on stock price.

W FORMATION

A pattern in which the price of a security or commodity reaches the support level on two occasions and then moves up again (Figure W.1). A reverse W formation is the opposite case: The price goes to a resistance level twice and then goes down.

FIGURE W.1—W FORMATION

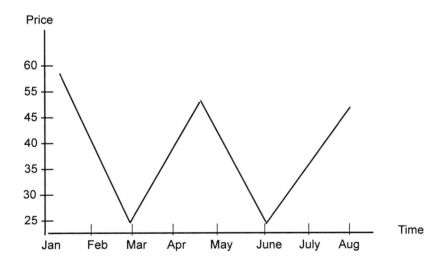

WALL STREET WEEK TECHNICAL MARKET INDEX

A consensus index of 10 technical stock market measures or indicators, developed by Robert Nurock. The index appears in *Futures* and *Investor's Analysis.* The index, updated each week, is used to spot intermediate to long-term market moves (three to six months or more) rather than short-term movements.

The overall index takes into account such factors as financial conditions, monetary factors, investor perceptions, speculation, insider purchases, market activity, and calls and puts. The index confirms that a trend will continue when most of the index components are neutral. The index warns of a possible change in a prevailing trend when at least five of the components are either negative or positive. Most importantly, the index is used to substantiate an upward or downward trend in the stock market, indicating if a market bottom or top has arrived. A market bottom is a buy indicator; a top says sell. Fundamental analysis is not taken into account, so ratios of profit, dividends, etc. are not considered.

The 10 component indicators comprising the *Wall Street Week* Technical Market Index are:

1. *Advisory Service Sentiment.* The extent to which stock market newsletters are positive, negative, or neutral on the stock market, with special attention to those newsletters predicting a market correction. Investors Intelligence collects the data. Advisory service sentiment is usually late in that it tends to follow the market trend rather than anticipate it. Hence, it is a contrary indicator. For example, if the services are bullish, the market will go down. The rule of thumb is that it is positive when the percentage of bears plus half the percentage anticipating a correction goes above 51.5%. This is an intermediate bottom indication. A reading below 35.5% is negative. This is an intermediate top indication.

2. The *Insider Activity Indicator* compares insider sell transactions to insider buy transactions. Since insiders are very knowledgeable, they are often correct in their investment decisions. This indicator is computed each week by Vickers Stock Research Corporation. A ratio of less than 1.42 means that insiders view their stocks as underpriced and expect them to go up. If the ratio exceeds 3.61, insiders consider their stocks overpriced and expect a price downturn.

3. The *Market Breadth Indicator* represents a moving total of the difference between advancing and declining issues over the previous

ten trading days. It reveals market strength or weakness by looking at the direction of stock prices and making comparisons to overall market indices. A Market Breadth Indicator that goes from below +1,000 to the point where it peaks out and drops 1,000 points from the peak is a positive reading. If the indicator contracts from above -1,000 to where it bottoms out and then increases 1,000 points from the trough, it is a negative reading. Neutral readings are from -1,000 to +1,000.

4. The *Dow Jones Momentum Ratio* looks at the percentage difference between the Dow Jones Industrial Average (DJIA) and its 30-day moving average. The ending DJIA for the last 30 days is totaled and then divided by 30. We then divide the DJIA's last close by its most recent 30-day average. If the DJIA diverges more than 3% from its 30-day average, the market is overbought; if the divergence is less than 3%, the market is oversold. A positive reading occurs if the DJIA is more than 3% below its moving average; the reading is negative if it is 3% above.

5. *Arm's Short-term Trading Index (TRIN)* is a 10-day moving average of the ratio of volume in advancing issues to that in declining issues on the New York Stock Exchange. The index measures the strength of volume. The index equals:

(Number of advancing issues/number of declining issues) divided by (volume of advancing issues/volume of declining issues).

A 10-day moving average is then computed. TRIN is a short-term measure of extremes in pessimism or optimism. A positive reading is above 1.20, a negative reading below .80. Anything in between is neutral.

6. The *New York Stock Exchange (NYSE) Hi-Low Index* looks at a moving average of the total number of stocks reaching new highs or new lows over the previous 10 trading days. At a market peak there should be few new lows; there should be few highs at a bottom. A reading is positive when there is an expansion of the 10-day average of new highs from less than 10 until it is more than the average number of new lows; the reading is negative when new lows move from fewer than 10 until more than the average number of new highs. A change in market direction is anticipated when either premise no longer holds.

7. *Prices of NYSE Stocks relative to Their Moving Averages:* This generates a positive reading if fewer than 30% of NYSE stocks are priced in excess of their 10-week moving average and fewer than

40% are priced in excess of their 30-week moving average. The reading is negative if more than 70% are above the 10-week average and more than 60% are above the 30-week average. Both the 10-week and 30-week moving averages must be met together for the measure to be reliable. This measure is used to warn of overvalued or undervalued stock prices. The data for this measure is compiled by Investors Intelligence.

8. The weekly *Option Premium Ratio* equals:

<u>Average premium on all listed put options</u>
Average premium on all listed call options

A ratio less than 42% means over-optimism, resulting in excessive call prices, a negative sign; a ratio over 95.5% shows too much pessimism, resulting in excessive put prices, a positive sign. The data are accumulated by the Options Clearing Corporation.

9. The *Low-Priced Activity Ratio* compares the trading volume in risky (speculative) stocks to that of blue chip Dow Jones Industrials. Speculative trading is typically high (above 7.59%) at a market peak and low (below 2.82%) at a market trough. A high ratio means a lot of speculation in the market, pointing to a possible downturn in stock prices. The information is published in Barron's each week.

10. *Federal Reserve Monetary Policy:* The weekly ratio of the federal funds rate to the federal discount rate attempts to evaluate the direction of Federal Reserve Board policy. If the Fed tightens the money supply, the Fed rate increases relative to the discount rate because the Fed is charging a higher rate between member banks and a lower rate (the discount rate) for member banks to borrow from the Fed. The ratio equals:

<u>Closing bid price for federal funds</u>
Daily discount rate

A four-business-day average of the ratio is computed. (Wednesday is not considered because it may be very volatile due to end-of-week bank transactions when banks establish reserve positions.) A positive reading is below 103%, a negative one above 125%. A low Federal funds rate compared to the discount rate implies an easy money policy, allowing business expansion, resulting in higher stock prices; a high rate implies tight money, business slowing, and lower stock prices.

These 10 technical measures are combined to obtain the *Wall Street Week* Technical Market Index.

Some guidelines for this index are:

- If the indicator is at an extreme, at a market top a minus is assigned, and a plus at the bottom.
- An indicator reading between extremes is neutral (0).
- The first time an indicator moves from one extreme to another in either direction, a neutral reading is assigned and maintained until it again reaches a plus or minus reading.

After this appraisal is done, the net total of the positive (+) and negative (-) indicators in the reading for the *Wall Street Week* Technical Market Index for the week is determined. The following is how the score is read:

Score	Meaning
-5	Extremely bearish; sell now
-4	Strongly bearish; get ready to sell
-3	Bearish
-2	Mildly bearish
-1	Neutral
0	Neutral
+1	Neutral
+2	Mildly bullish
+3	Bullish
+4	Strongly bullish; get ready to buy
+5 or higher	Extremely bullish; buy now

The index can range from -10 to +10.

When the reading is +5 or better, buy stocks immediately.

WEAK MARKET

A weak market is characterized by a preponderance of sellers over buyers and a declining price trend.

See also Strong Market.

WEDGE

A technical chart pattern similar to but slightly different from a triangle pattern (see Figure W.2). A wedge is formed by two converging lines that move in the same direction, connecting a series of peaks and troughs. An increasing wedge interrupts a declining price trend, a decreasing wedge interrupts an uptrend.

FIGURE W.2—WEDGE

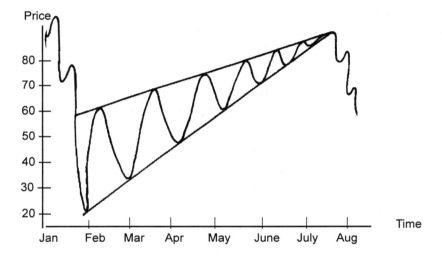

WHIPSAW

Whipsaw occurs when a trader's technical indicators have many buy and sell signals in a narrow price range that do not move in the predicted direction, resulting in significant trading losses.

WILDER'S AVERAGE TRUE RANGE

Welles Wilder formulated the average true range (ATR), which measures the degree of variability in the market. The wider the range, the greater the volatility.

At a market bottom, there is likely to be a high ATR. At a market top, the ATR is likely to be low. The market bottom may come after panic selling, the market top after a consolidation period or rally.

ATR is based on a moving average (e.g., 15 days) of the true ranges. The true range is the greatest of either:

- The difference between yesterday's close and today's low
- The difference between yesterday's close and today's high
- The difference between today's high and today's low

WILLIAMS' %R

Developed by Larry Williams, Williams' %R (pronounced "percent R") is a momentum indicator that measures overbought/oversold levels. Williams' %R is very similar to the Stochastic Oscillator, except that %R is plotted upside-down and the Stochastic Oscillator has internal smoothing. To display the Williams' %R indicator on an upside-down scale, it is usually plotted using negative values (e.g., -20 percent). We simply ignore the negative symbols.

The rule is:

- Readings in the range of 80% to 100% indicate that the security is oversold
- Readings in the 0% to 20% range suggest that it is overbought.

As with all overbought/oversold indicators, it is best to wait for the price to change direction before making your trades. For example, if an overbought/oversold indicator is showing an overbought condition, it is wise to wait for the security's price to move down before selling the security. (The MACD is a good indicator to monitor change in a security's price.) It is not unusual for these indicators to remain overbought or oversold for a long time as the security's price continues to climb or fall. Selling simply because the security appears overbought may take you out of the security long before its price shows signs of deterioration. An interesting phenomenon of the Williams' %R indicator is its uncanny ability to anticipate a reversal in the security price. The indicator almost always forms a peak and turns down a few days before the security's price peaks and turns down; and thus %R usually creates a trough and turns up a few days before the price turns up.

Williams' %R is calculated as:

$$\frac{\text{Highest High in n-periods - Today's Close}}{\text{Highest High in n-periods - Lowest Low in n-periods}} \times (-100)$$

YIELD CURVE

While basically a graphic representation of bond yields versus bond maturities, the yield curve is often quoted by analysts to describe the condition of the bond market. The curve breaks bonds into three categories by maturity: short (less than 5 years long), intermediate (5 to 10 years), and long (more than 10 years).

The curve is drawn by graphing the maturities of one type of bond from shortest to longest (typically from overnight to 30 years) against the yields those bond maturities are currently producing. The curve is then sometimes compared to yield curves of other securities, or of the same security at earlier times.

The yield curve can be found daily in *Investor's Business Daily* and the *Wall Street Journal*. Many other publications and newsletters, such as the *Bond Fund Report* or *Grant's Interest Rate Observer,* as well as reports from investment houses typically discuss and chart the yield curve. The Bloomberg Website, for example, presents the Treasury yield curve.

The slope of the yield curve and relative changes in its shape can help investors understand bond market conditions.

When the yield curve is steep—short-term rates low, long-term rates high, which is considered the normal shape—to boost their returns investors must take on the added risk of owning long-term bonds. Long-term bond prices are most volatile because their fixed payouts can be dramatically hurt if, for instance, inflation were to rise and erode buying power.

An inverted yield curve creates a difficult choice for investors. Should lower long-term rates be locked in? Or can they take the chance that the currently more attractive short-term rates will stay? This often occurs in times of economic difficulty.

A flat yield curve, when passbook rates might be the same as mortgage rates, is often seen as a signal of an upcoming dramatic change in the interest rate environment.

Caution: Yield is not the only consideration in buying a bond. The price risk of holding long-term bonds should be studied. Credit quality is another important factor—a high current yield is not valuable after an issuer defaults.

YIELD GAP

A stock/bond yield gap measures the difference in yield between securities. Normally, it measures yield spread between the average quarterly yield on long-term corporate bonds and the quarterly dividend yield on industrial stocks.

Stock yields are usually higher than bond yields, given the greater risk. If bond yields are higher, the yield gap becomes negative and is known as the "reverse" yield gap. This negative gap is a direct result of growing investor confidence in stocks; they are considering stocks less risky.

This gap is viewed as a long-term indicator of market valuation. Investors need to detect the long-term trend for the gap. For example, if the gap is negative (i.e., bond yields are higher than stock yields) for some time, it is usually indicating a bull market in equities. A positive gap is normally associated with a bear market.

Z

ZIGZAG CHART

The zigzag indicator presents major (not insignificant) up or down changes when plotting the price of a security. The final "leg" shown in a zigzag chart can change due to changes in the underlying plot. A change in stock price can alter a previous value of the indicator. In other words, the zigzag measure can modify its values depending on future changes in the underlying plot. There is hindsight into past prices. A strong advantage with zigzag analysis is to spot major reversals in stock prices.

The chart is formed by putting imaginary points where prices reversed by a minimum specified amount. The points are then connected by a straight line.

APPENDIX A

SOURCES OF TECHNICAL DATA

Sources of technical information useful in analysis include:

FINANCIAL PUBLICATIONS

Barron's (weekly)
Dow Jones and Company
200 Burnett Road
Chicopee, MA 01021

Futures: The Magazine of Commodities and Options (monthly)
219 Parkade
Cedar Falls, IA 50613

Investor's Business Daily (daily)
12655 Beatrice Street
Los Angeles, CA 90066

Technical Analysis of Stocks and Commodities (monthly)
3517 S.W. Alaska Street
Seattle, WA 98126

Wall Street Journal (daily)
Dow Jones and Company
200 Burnett Road
Chicopee, MA 01021

ECONOMETRIC DATA

Institute for Econometric Research
3471 North Federal Highway
Fort Lauderdale, FL 33306

INVESTMENT ADVISORY REPORTS AND INFORMATION

Lowry's Reports (weekly)
701 North Federal Highway
North Palm Beach, FL 33408

Offers buy and sell advice based on calculations of buying power and selling pressure. The reports are based on changes in stock prices and the associated volume.

Merrill Analysis, Inc.
Box 228
Chappaqua, NY 10514

Zweig Forecast
P.O. Box 5345
New York, NY 10150

MARKET INFORMATION

Investment Company Institute
1600 M Street N.W.
Washington, DC 20036

Investors Intelligence
30 Church Street
New Rochelle, NY 10801

Vickers Stock Research Corporation
Box 59
Brookside, NJ 07926

APPENDIX B

SUBSCRIPTION CHARTING SERVICES

Chartcraft
30 Church Street
New Rochelle, NY 10801

Stocks, options, commodities and financial futures, and technical indicators.

Commodity Price Charts
219 Parkade
Cedar Falls, IA 50613

Financial futures and commodities.

Commodity Research Bureau
Suite 1820
305 Wacker Drive
Chicago, IL 60606

Commodities and financial futures.

Daily Graphs
P.O. Box 24933
Los Angeles, CA 90024

Options and stocks.

Knight-Ridder Commodity Research Bureau
Suite 1850
100 Church Street
New York, NY 10007

Commodities and financial futures.

Standard and Poor's
25 Broadway
New York, NY 10024

Stocks.

INDEX